Applied Mineralogy Handbook

Applied Mineralogy Handbook

Editor: Trinity Collins

www.callistoreference.com

Callisto Reference,
118-35 Queens Blvd., Suite 400,
Forest Hills, NY 11375, USA

Visit us on the World Wide Web at:
www.callistoreference.com

ISBN: 978-1-64116-792-5 (Hardback)

Cataloging-in-publication Data

Applied mineralogy handbook / edited by Trinity Collins.
p. cm.
Includes bibliographical references and index.
ISBN 978-1-64116-792-5
1. Mineralogy. 2. Minerals. I. Collins, Trinity.
QE363.2 .M56 2023
549--dc23

Table of Contents

Preface

In my initial years as a student, I used to run to the library at every possible instance to grab a book and learn something new. Books were my primary source of knowledge and I would not have come such a long way without all that I learnt from them. Thus, when I was approached to edit this book; I became understandably nostalgic. It was an absolute honor to be considered worthy of guiding the current generation as well as those to come. I put all my knowledge and hard work into making this book most beneficial for its readers.

Mineralogy refers to the systematic study of all minerals, including their description, physical environment, crystallography and chemical characteristics. Applied mineralogy is focused on inorganic non-metal materials, and their sampling, analysis, synthesis, preparation, evaluation and property identification. Process mineralogy is a type of applied mineralogy that concentrates on solving issues related to ore processing. The goal of process mineralogy is to recognize, diagnose and forecast the mineralogically affected or regulated processing properties of an ore. They are also used as essential components of building materials like marble, plaster, granite, limestone, glass, cement and gravel. This book includes some of the vital pieces of work being conducted across the world, on various topics related to applied mineralogy. It will serve as a valuable source of reference for graduate and postgraduate students. Most of the topics introduced in this book cover new techniques and the applications of this field.

I wish to thank my publisher for supporting me at every step. I would also like to thank all the authors who have contributed their researches in this book. I hope this book will be a valuable contribution to the progress of the field.

Editor

Advanced Materials with Improved Characteristics, Including Technical Ceramics and Glass

1

Phase Changes in Radiation Protection Composite Materials Based on Bismuth Oxide

S. Yashkina(✉), V. Doroganov, E. Evtushenko, O. Gavshina, and E. Sysa

Department of Technology of Glass and Ceramics, Belgorod State Technological University named after V.G. Shukhov, Belgorod, Russia
yashkina_asp@mail.ru

Abstract. XRF method was used to study the mineralogical compounds of the radiation protection ceramic materials based on a high-alumina binder and a "heavy" aggregate, bismuth oxide. The content of Bi_2O_3 in the test samples was kept in the range of 38.5–75 wt%. Along with the bismuth oxide, the aggregate was aluminum oxide (Al_2O_3). The binder synthesis followed the principle of obtaining ceramic concretes based on artificial ceramic binders (ACB).

The paper establishes the specifics of sintering of the composites under study, fired under different temperatures.

Keywords: Bismuth oxide · Artificial ceramic binders · X-ray fluorescence · Radiation protection ceramic composites

1 Introduction

Bismuth oxide is a widely used bismuth compound due to its chemical stability, low toxicity, and its unique physical and chemical properties (Gulbin et al. 2014). Its sufficiently high density (8.9 g/cm^3) allows to use it in the synthesis of "heavy" materials employed in the protection from gamma rays. As it is known, in this case, the leading role in characterizing the protective properties is played by the density of the material. In the way of exercising high structural properties, the most efficient were the ceramic matrices (Correya et al. 2018).

This paper studies the phase transformations of the radiation protection materials based on the high-alumina ACBs, bismuth oxide, and aluminum oxide sintered at different temperatures.

2 Methods and Approaches

Obtaining the artificial ceramic binder (ACB) was carried out through the technique of wet grinding in a periodic action ball mill with a stepwise loading of material. This principle of obtaining ceramic concretes allows to obtain matrices with the desired phase composition, improved physical and mathematical properties and ensures the broad opportunities for using various aggregates.

The suspension based on high-alumina grog possesses the thixotropic dilatant flow (the destruction of the initial thixotropic structure and the following dilatant formation of structure). The bismuth oxide powder used as aggregate represents spherical particles sized no more than 35 μm.

The source materials were mixed in the proportions presented in Table 1. Then, the technique of static pressing was used with the specific pressure of 100 MPa to form the bar. After drying, the samples were subjected to burning within the interval of 200° to 1000 °C with holding at the maximum temperature for one hour.

Table 1. Compositions of radiation protection materials

Number of the compound	Content, wt%		
	ACB	Al_2O_3	Bi_2O_3
1	38.5	23.0	38.5
2	19.0	23.0	58.0
3	–	25.0	75.0

The X-ray fluorescence analysis was performed using the DRON-3 diffractometer. XRD patterns were recorded using the CuKα radiation (Ni filter); tube voltage: 20 kV; tube anode current: 20 mA; measuring range: 10000–4000 PPS; detector turn rate: 2.4 rpm; elevation angle: 10. The JCPDF database was used for identification.

3 Results and Discussion

The analysis of the transformations of the samples with the admixture Bi_2O_3 at varied temperatures (Figs. 1, 2 and 3) indicates that heating within the interval of 100–500 °C, all compounds are mineralogically constant and exhibit phases of mullite, corundum, and Bi_2O_3.

When increasing the burning temperature to 600 °C, starts the formation of bismuth silicates $2Bi_2O_3{\cdot}3SiO_2$ and $12Bi_2O_3{\cdot}2SiO_2$. Further increase of the burning temperature to 700 °C leads to a more intensive formation of bismuth silicates, indicated by the increase in the intensity of reflections for these compounds. In the process, the entire Bi_2O_3 turns into silicates, while the interplanar reflections of bismuth oxide totally disappear. There occurs a sharp decrease in the intensity of mullite peaks, while in the composition No. 3 this compound is totally absent.

When the temperature is increased up to 800 °C and further, all compositions display further formation of bismuth silicates, and in the compositions No. 2–3 there were identified two phases of α-Al_2O_3 и $12Bi_2O_3{\cdot}2SiO_2$, while in the composition No. 1, in addition to the indicated two, there also forms $2Bi_2O_3{\cdot}3SiO_2$ (Figs. 1, 2 and 3).

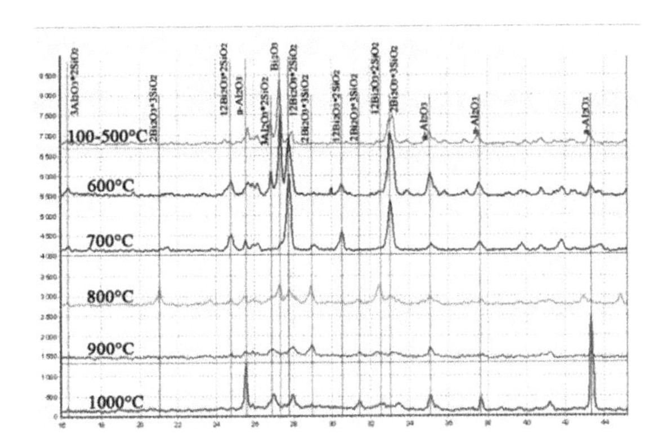

Fig. 1. X-ray patterns of the samples of the composition of radiation protection composites No. 1

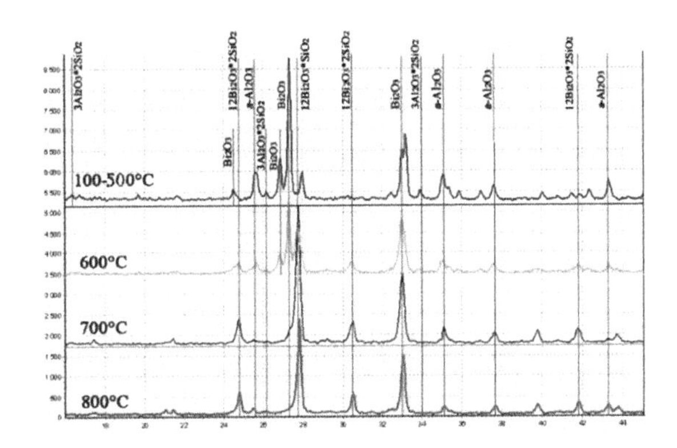

Fig. 2. X-ray patterns of the samples of the composition of radiation protection composites No. 2

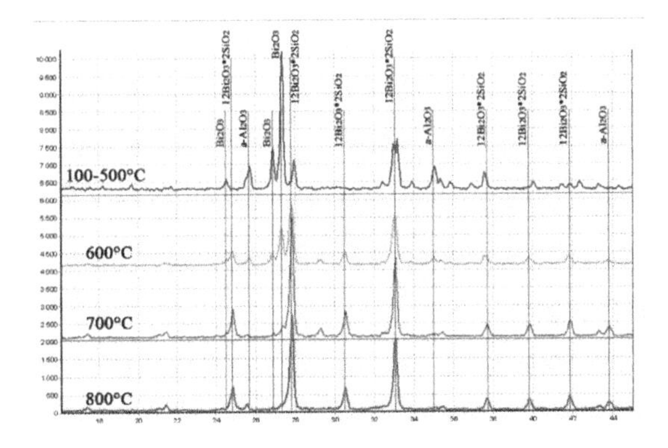

Fig. 3. X-ray patterns of the samples of the composition of radiation protection composites No. 3

It is possible to assume that compositions No. 1–2 show the formation of bismuth silicates as a result of mullite decomposition and binding of SiO_2 into the bismuth silicates, while in the composition No. 3 the silicates form due to the interaction of Bi_2O_3 and the nanosilica introduced as binder.

4 Conclusions

This paper studied the mineralogical compounds of the radiation protection ceramic materials based on a high-alumina ACB and bismuth oxide. It was identified that the higher the content of Bi_2O_3 in the source composition, the more intensive is the process of mullite decomposition. This way, at the content of Bi_2O_3 38,5%, mullite perseveres up to the temperature of 900 °C, and if bismuth oxide is increased to 58%, mullite disintegrates completely at the temperature of 700 °C, which is 200 °C lower than the previous composition.

Acknowledgement. The work is realized in the framework of the Program of Flagship University development on the base of the Belgorod State Technological University named after V.G. Shukhov, using the equipment of High Technology Center at BSTU named after V.G. Shukhov.

References

Correya AA, Mathew S, Nampoori VPN, Mujeeb A (2018) Structural and optical characterization of hexagonal nanocrystalline bismuth-bismuth oxide core-shell structures synthesized at low temperature. Optik 175:930–935

Gulbin VN, Kolpakov NS, Polivkin VV (2014) Radio - i radiatcionno-zashchitnye kompozitcionnye materialy s nanostrukturnymi napolniteliami [Radiation protection composite materials with nanostructure aggregates]. VSTU Bull 23:43–51

2

Experimental Modeling of Biogeosorbents

T. Shchemelinina[1], O. Kotova[2], E. Anchugova[1], D. Shushkov[2(✉)], G. Ignatyev[2], and M. Markarova[2]

[1] Laboratory of Biochemistry and Biotechnology, Institute of Biology, Komi Science Center, Ural Branch of the Russian Academy of Sciences, Syktyvkar, Russia

[2] Laboratory of Technology of Mineral Raw, Institute of Geology named after Academician N.P. Yushkin Komi Science Center of the Ural Branch of the Russian Academy of Sciences, Syktyvkar, Russia

dashushkov@geo.komisc.ru

Abstract. A new trend in the modeling algorithm oil sorption materials is the adsorptive immobilization of strains of microorganisms on mineral sorbent. The objects of study were clay and zeolite raw and biogeosorbents with oil-oxidizing microorganisms from Biotrin preparation, immobilized on them. During our work we modeled biogeosorbents and estimated their sorption and destructive properties in reference to petroleum hydrocarbons.

Keywords: Biogeosorbents · Biotrin · Petroleum hydrocarbons · Sorption · Zeolites · Clays

1 Introduction

One of the most promising ways to solve problem of oil pollution is to use biogeosorbents, which are sorbent-carriers with oil-oxidizing microorganisms immobilized on their surface (Shchemelinina et al. 2017). It is possible to use zeolite and clay raw as mineral carriers, which have high ion-exchange and sorption properties (Kotova et al. 2017). The purpose of this work is to study sorption and destructive properties of biogeosorbents based on clay and zeolite raw as mineral carriers for Biotrin biopreparation.

2 Methods and Approaches

The objects of study were:

1. Mineral carriers based on analcime-bearing rocks from Koinskaya zeolite area (Shushkov et al. 2006), clinoptilolite-bearing clays and glauconite-bearing rocks from Chim-Loptyugskoe oil shale deposit (Saldin et al. 2013; Simakova 2016), located in Komi Republic (Russia). For comparison, Ionsorb ™ quartz-glauconite sand from Bondarskoe deposit of Tambov region was taken as a control.

2. Strains of microorganisms in the composition of Biotrin biopreparation (Conclusion ... 2017): bacteria *Pseudomonas yamanorum* VKM B-3033D, isolated from heavily soiled railway bed near the city of Syktyvkar (Patent 2615458 RU); yeast *Rhodotorula glutinis* VKM Y-2998D (Patent 2658134 RU); microalgae *Chlorella vulgaris* Beijer. f. *globosa* V. Andr. A1123. Microorganisms (cell titer 10^9) were cultivated according to standard methods. Immobilization of the biopreparation on mineral carriers was carried out in the ratio of 1 part of the biopreparation to 6 parts of the sorbent. Initial sorbents (without Biotrin) and biogeosorbents were added to oil-contaminated water, aerated for 4 days, and total petroleum hydrocarbons (TPH) content in water samples, filtered initial sorbents and biogeosorbents was measured (Method ... 1998).

3 Results and Discussion

Norms for maximum permissible concentration (MPC) of TPH in water of fishery value are 0.05 mg/dm^3 (Order ... 2016). The TPH content in the control water sample is 2.4 times higher than MPC (Table 1).

Experiments showed that samples of the initial analcime-bearing rocks (551, 56403, 1/83) presented adsorption activity in relation to oil products. As a result of the introduction of these samples into oil-polluted water, the content of pollutant in water decreases 2.5–3 times in 4 days, to the MPC. When biogeosorbents are applied to contaminated water (551-B, 56403-B, 1/83-B), the efficiency of water purification decreases and does not reach MPC standards, which indicates decreasing sorption properties after immobilization of microorganisms on the mineral carriers. This is probably due to decreasing surface area of the mineral carriers covered by microorganisms. During the study of initial samples and Biotrin treated samples for destructive properties we revealed that the efficiency of oxidation of oil products in samples 551-B, 56403-B, 1/83-B increases in 4.4, 3.5 and 1.14 times, respectively.

The sorption properties of clinoptilolite-bearing clays are most attractive in sample 541-31. However, taking into account a highly destructive activity of microorganisms in 538-35-B biogeosorbent, it is preferred for remediation of oil contaminated water.

Samples of initial glauconite-bearing rocks have high sorption properties (539-40, 531-56, 315-10, TG). The TPH content in the experimental water is reduced by 3.4–5 times in 4 days relative to a control sample. Biodegradation of hydrocarbons in samples of Biotrin immobilized glauconite-bearing rocks (539-40-B, 531-56-B, 315-10-B, TG-B) ranges from 62 to 76%.

Table 1. Change in the concentration of oil products in water in the presence of initial mineral carriers and biogeosorbents

Initial samples	TPH content*	Biotrin treated samples	TPH content*
Analcime-bearing rocks			
551	$\frac{0.04 \pm 0.014}{250 \pm 60}$	551-B	$\frac{0.11 \pm 0.04}{57 \pm 23}$
56403	$\frac{0.046 \pm 0.016}{130 \pm 50}$	56403-B	$\frac{0.061 \pm 0.021}{37 \pm 15}$
1/83	$\frac{0.048 \pm 0.017}{250 \pm 60}$	1/83-B	$\frac{0.071 \pm 0.025}{220 \pm 90}$
58603	$\frac{0.071 \pm 0.025}{250 \pm 60}$	58603-B	$\frac{0.064 \pm 0.022}{90 \pm 40}$
Clinoptilolite-bearing clays			
538-35	$\frac{0.085 \pm 0.030}{50 \pm 20}$	538-35-B	$\frac{0.037 \pm 0.013}{40 \pm 16}$
541-31	$\frac{0.035 \pm 0.012}{250 \pm 60}$	541-31-B	$\frac{0.058 \pm 0.021}{100 \pm 40}$
Glauconite-bearing rocks			
539-40	$\frac{0.024 \pm 0.009}{58 \pm 23}$	539-40-B	$\frac{0.072 \pm 0.025}{20 \pm 8}$
531-56	$\frac{0.027 \pm 0.009}{63 \pm 25}$	531-56-B	$\frac{**}{15 \pm 6}$
315-10	$\frac{0.09 \pm 0.03}{11 \pm 4}$	315-10-B	$\frac{0.021 \pm 0.007}{17 \pm 7}$
TG	$\frac{0.035 \pm 0.012}{90 \pm 40}$	TG-B	$\frac{0.045 \pm 0.016}{34 \pm 14}$
Oil-contaminated water (control)	0.12 ± 0.041		

Note: * – in the numerator, TPH content in the experimental water, mg/dm^3, in the denominator – TPH content in the initial sorbents and biogeosorbents after the experiment, mg/g. ** – no data

4 Conclusions

Our experiments resulted in modeling of biogeosorbents based on clay and zeolite raw and oil-oxidizing microorganisms from Biotrin biopreparation immobilized on them. We determined that samples of initial sorbents showed a high adsorption activity with respect to oil products. TPH content in water was reduced by 2.5–5 times, up to or substantially below MPC. We revealed that microorganism cells could reduce sorption properties of mineral carriers, at the same time providing oil destruction. Biodestruction of oil products with biogeosorbents for 4 days was 12–77%.

Acknowledgments. The authors express their gratitude to the Center for Collective Use "Geonauka", ecoanalytical laboratory of the Institute of Biology of the Komi Science Center UB RAS for their assistance in analytical work. The work was carried out with the partial financial support of UB RAS Programs (project 18-5-5-44), UMNIK project (12412GU/2017), of the State task "Development of biocatalytic systems based on enzymes, microorganisms and plant cells, their immobilized forms and associations for the processing of plant raw, production of biologically active substances, biofuels, remediation of contaminated soils and wastewater treatment" No. AAAA-A17-117121270025-1, "Scientific basis for effective subsoil use, development and exploration of mineral resource base, development and implementation of innovating technologies and economic zoning of the Timan-Nothern Ural region" No. AAAA-A17-117121270037-4.

References

Certificate on the toxicological and hygienic assessment of "BIOTRIN" consortium of strains of oil-oxidizing microorganisms. State federal enterprise for science Research center for toxicology and hygienic regulation of biopreparations at Federal medico-biological agency, Serpukhov, 28 September 2017 (in Russian)

(1998) Method for performing measurements of the mass fraction of petroleum products in soil samples on a Fluorat-02 analyzer. Institute of Biology of Komi Science Center of the Ural Branch of the Russian Academy of Sciences, 16.1.21-98, Moscow, 15 p (in Russian)

Patent of the Russian Federation No. 2615458

Patent of the Russian Federation No. 2658134

Order of the Ministry of Agriculture of the Russian Federation of December 13, 2016 No. 552. On approval of water quality standards for water bodies of fishery importance, including standards for maximum permissible concentrations of harmful substances in the waters of water bodies of fish-economic importance (in Russian)

Saldin VA, Burtsev IN, Mashin DO, Shebolkin DN, Inkina NS (2013) Marking horizons in the Upper Jurassic deposits of the Yarengsky shale region (north-east of the Russian plate), no 11, pp 26–29. Vestnik of the Institute of Geology of the Komi Science Center UB RAS (in Russian)

Simakova YuS (2016) Features of globular layered silicates of the Chim-Loptyugsky oil shale deposit, № 9–10, pp 52–57. Vestnik of the Institute of Geology of the Komi Science Center UB RAS (in Russian)

Shushkov DA, Kotova OB, Kapitanov VM, Ignatiev AN (2006) Analcime rocks of Timan as a promising type of minerals. Syktyvkar, 40 p (Scientific recommendations to the national economy/Komi Science Center UB RAS, issue 123) (in Russian)

Shchemelinina TN, Kotova OB, Harja M, Anchugova EM, Pelovski I, Kretesku I (2017) New trends in the mechanisms of increasing the productivity of materials on a mineral basis, no 6, pp 40–42. Vestnik of the Institute of Geology of the Komi Science Center UB RAS (in Russian)

Kotova OB, Harja M, Cretescu I, Noli F, Pelovski Y, Shushkov DA (2017) Zeolites in technologies of pollution prevention and remediation of aquatic systems, no 5, pp 49–53. Vestnik of the Institute of Geology of the Komi Science Center UB RAS

3

Three-Cation Scandium Borates RxLa1 –xSc3(BO3)4(R = Sm, Tb): Synthesis, Structure, Crystal Growth and Luminescent Properties

A. Kokh[1(✉)], A. Kuznetsov[1], K. Kokh[1,2], N. Kononova[1], V. Shevchenko[1], B. Uralbekov[3], A. Bolatov[3], and V. Svetlichnyi[4]

[1] Sobolev Institute of Geology and Mineralogy SB RAS, Novosibirsk, Russia
a.e.kokh@gmail.com
[2] Novosibirsk State University, Novosibirsk, Russia
[3] Al-Farabi Kazakh National University, Almaty, Kazakhstan
[4] Tomsk State University, Tomsk, Russia

Abstract. Complex ortohoborates of rare earth metals with the general chemical formula $R_xLa_{1-x}Sc_3(BO_3)_4$ (R = Sm, Tb) have been obtained by solid state synthesis and spontaneous crystallization. These crystals belong to the huntite family with the space group R32 and for x = 0.5 have unit cell parameters a = 9.823(6), c = 7.975(3) (SLSB) and a = 9.803(3), c = 7.960(4) Å (TLSB).

Keywords: Crystal · Borate · Structure · Huntite · Growth · Luminescence

1 Introduction

Orthoborates with the general formula $RX_3(BO_3)_4$, where R = Y, Ln; X = Al, Ga, Sc, Cr, Fe are practically important and interesting from the point of view of crystal chemistry objects for research. One of the important properties of these compounds is the ability to form a non-centrosymmetric structure, which is called huntite-like. Such a structure causes, for example, non-linear optical properties.

To understand the formation of the huntite-like structure of three-cation scandoborates, we consider the lanthanum – scandium borate $LaSc_3(BO_3)_4$. The authors (He et al. 1999) distinguish three modifications of this crystal: high-temperature monoclinic with the C2/c space group, medium temperature trigonal with the R32 space group (huntite-like) and low-temperature monoclinic with the Cc space group. As a result of our research (Fedorova et al. 2013) identity of the X-ray patterns of polymorphic modifications high and low was shown.

The stabilization of the huntite-like structure can occur if an additional isomorphic cation is introduced into the $LaSc_3(BO_3)_4$ structure, that was confirmed in (Li et al. 2001) who initiated the new three-cation scandoborate with the huntite-like structure $Nd_xLa_{1-x}Sc_3(BO_3)_4$. Further, in a number of works by adding a third cation

$R_xLa_ySc_z(BO_3)_4$ nonlinear optical crystals with a stable huntite-like structure were obtained with R = Gd (Xu et al., 2011); Y (Ye et al. 2005) and Lu (Li et al. 2007).

The existence of a huntite-like structure for the boundary members of the REE series suggests the stability of such a structure with the rest of the REE. This paper presents data on the huntite-like structures SLSB and TLSB for systems $R_xLa_{1-x}Sc_3(BO_3)_4$ (R = Sm, Tb).

2 Methods and Approaches

Polycrystalline sample of $R_xLa_{1-x}Sc_3(BO_3)_4$ (x = 0–0.5)were prepared by the method of two stage solid state synthesis in a Pt crucible. The stoichiometric mixtures of pure raw La_2O_3, Sc_2O_3, H_3BO_3 and R_2O_3(R = Sm, Tb) reactants were heated at 800 °C for 5 h to decompose H_3BO_3. At the second stage, the mixtures were grinded in an agate mortar and heated again at 1300 °C for 12 h until the powder X-ray method showed no peaks of initial compounds (Fig. 1).

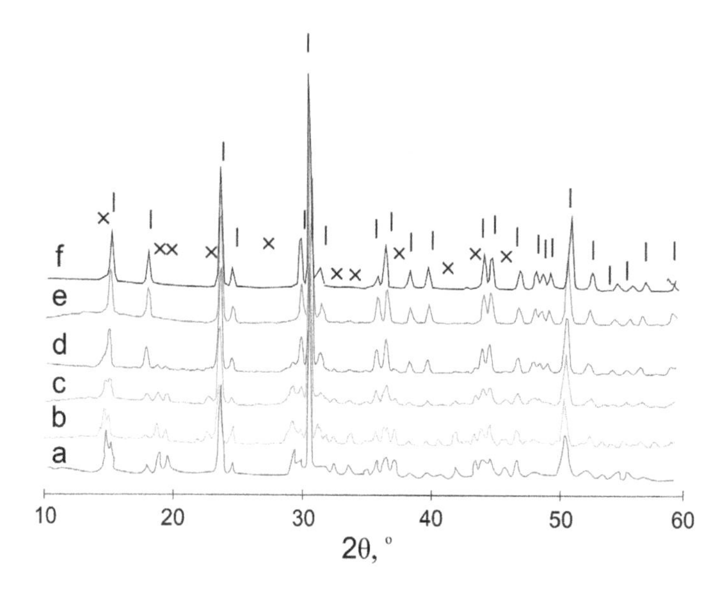

Fig. 1. X-ray pattern of $Sm^xLa^{1-x}Sc^3(BO^3)^4$, where x = 0(a), 0.2(b), 0.3(c), 0.4(d), 0.5(e), 1(f). I- huntite (R32), X- monoclinic (C2/c) structure.

Spontaneous crystals of $R_xLa_{1-x}Sc_3(BO_3)_4$with dimensions 30 × 30 × 10 mm with a transparent area of 5 × 5 × 5 mm were grown from $LiBO_2$- LiF flux Fig. 2. A Pt crucible containing $R_{0.5}La_{0.5}Sc_3(BO_3)_4$, Li_2CO_3, H_3BO_3 and LiF in the molar ratio of 1:1,5:1,5:3 was heated to 1000 °C. The charge was held in a melted state for a day to achieve homogenization. After this stage a platinum wire with a loop was placed in the center of the melt surface and the temperature was decreased to 900 °C. Then the melt was cooled with the 2 °C/day to 850 °C and following cooling at the rate of

15 °C/day to room temperature. The crystal was chosen for x-ray analysis. Powder diffraction patterns were refined using the Rietveld method within the GSAS- II program.

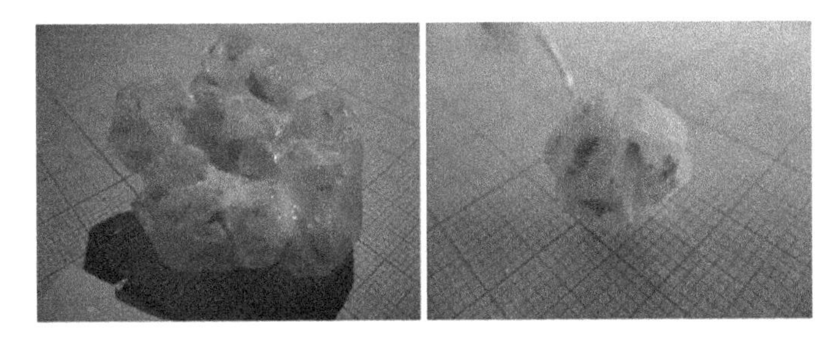

Fig. 2. Crystals grown from $LiBO_2$-LiF flux: TLSB (left) and SLSB (right)

The chemical composition of obtained crystals was measured by X-ray fluorescent analysis using XRF 1800 (Shimadzu, Japan). The results of the analysis are conformed with the formula obtained after crystal structure refinement: (Table 1)

Table 1. Composition of TLSB and SLSB based on X-ray fluorescent elemental analysis

Composition of the $Tb_xLa_ySc_z(BO_3)_4$		Ratio of Tb/La
Starting melt	$Tb_{0.5}La_{0.5}Sc_3(BO_3)_4$	1
Center	$Tb_{0.22}La_{0.78}Sc_3(BO_3)_4$	0.28
Edge	$Tb_{0.24}La_{0.75}Sc_{2,99}(BO_3)_4$	0.32
Composition of the $Sm_xLa_ySc_z(BO_3)_4$		Ratio of Sm/La
Starting melt	$Sm_{0.5}La_{0.5}Sc_3(BO_3)_4$	1
Center	$Sm_{0.32}La_{0.69}Sc_{2,98}(BO_3)_4$	0.46
Edge	$Sm_{0.35}La_{0.68}Sc_{2,97}(BO_3)_4$	0.52

3 Results and Discussion

Structure. According to Rietveld refinement both SLSB and TLSB crystalize in the trigonal space group R32 with unit cell parameters: a = 9.823(6), c = 7.975(3) (SLSB) and a = 9.803(3), c = 7.960(4)Å (TLSB). The structure framework is composed of the R, La atoms, Sc atoms and B atoms occupy trigonal prisms, octahedra and planar triangle of oxygen, respectively. The isolated (R, La)O_6 trigonal prisms alternate along the c-axis with BO3 triangle that are perpendicular to the c-axis. ScO_6octahedra link to each other along the edge and form twisted chain along c, which separate (R, La)O_6 prisms as well. The discrepancies between refined diffraction spectra with model calculations can be explained by crystal cleavage along {202} and {113}.

Luminescence. Typical excitation and luminescence spectra of SLSB are shown on Fig. 3(a). The strongest excitation peak of samarium crystal corresponds to $^6H_{5/2} \rightarrow {}^4F_{7/2}$ transition located at 407 nm. Whereas luminescent spectrum of SLSB has some peaks corresponding to$^4G_{5/2} \rightarrow {}^6H_J$ (J = 5/2, 7/2, 9/2 и 11/2) and located at 566, 602, 645 and 708 nm.

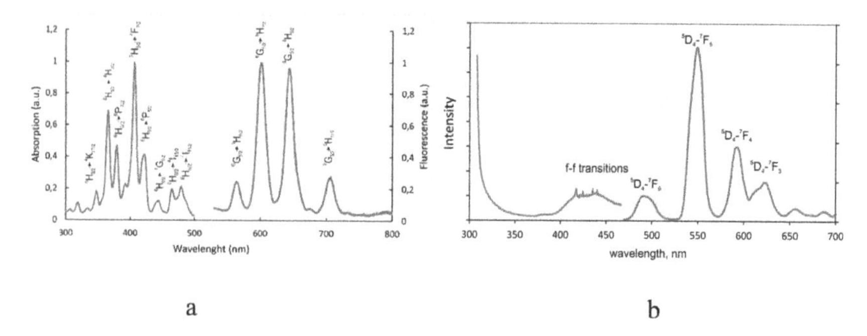

Fig. 3. Luminescent properties of (a) SLSB, (b) TLSB

On Fig. 3(b) TLSB excitation and luminescence spectra are shown with a wide strip at 300 nm corresponding to^4F - ^5D transition. Luminescent spectra consist of 5 peaks at 490, 505, 585 and 640 corresponding to$^5D_4$ - ^7F_J (J = 6, 5, 4, 3) transitions.

4 Conclusions

The formation of a huntite structure in systems $R_xLa_{1-x}Sc_3(BO_3)_4$, (R = Sm, Tb), as well as the dependence of the compositions stable in the required structure depending on the production method is shown. The spectral characteristics confirm the potential of using crystals as luminescent materials.

Acknowledgements. This work is supported by RFBR project#19-05-00198a, state contract of IGM SB RAS and partially by Project GF MES RK IRN AP05130794.

References

Fedorova MV, Kononova NG, Kokh AE, Shevchenko VS (2013) Growth of MBO3 (M = La, Y, Sc) and $LaSc_3(BO_3)_4$ Crystals from $LiBO_2$–LiF Fluxes. Inorg Mater 49:482–486

He M, Wang G, Lin Z et al (1999) Structure of medium temperature phase β-$LaSc_3(BO_3)_4$ crystal. Mater Res Innov 2(6):345–348

Li W, Huang L, Zhang G, Ye N (2007) Growth and characterization of nonlinear optical crystal $Lu_{0.66}La_{0.95}Sc_{2.39}(BO_3)_4$. J Cryst Growth 307:405–409

Li Y, Aka G, Kahn-Harari A, Vivien D (2001) Phase transition, growth, and optical properties of $NdxLa_{1-}xSc_3(BO_3)_4$ crystals. J Mater Res 16:38–44

Xu X, Ye N (2011) $Gd_xLa_{1-x}Sc_3(BO_3)_4$: a new nonlinear optical crystal. J Cryst Growth 324:304–308

Ye N, Stone-Sundberg JL, Hruschka MA et al (2005) Nonlinear Optical Crystal $Y_xLa_ySc_z(BO_3)_4$ $(x + y + z = 4)$. Chem Mater 17:2687–2692

4

Efficiency Evaluation for Titanium Dioxide-Based Advanced Materials in Water Treatment

M. Harja[1], O. Kotova[2], S. Sun[3], A. Ponaryadov[2](✉), and T. Shchemelinina[4]

[1] Gheorghe Asachi Technical University of Iasi, Iaşi, Romania
[2] Institute of Geology Komi SC UB RAS, Syktyvkar, Russia
avponaryadov@geo.komisc.ru
[3] Institute of Non-metallic Minerals, Department of Geological Engineering, School of Environment and Resource, Southwest University of Science and Technology, Mianyang, People's Republic of China
[4] Institute of Biology Komi SC UB RAS, Syktyvkar, Russia

Abstract. We present a comparative evaluation of efficiency of titanium dioxide polymorphs as an active photocatalyst (commercially available DegussaP25, anatase (Sigma Aldrich), natural leucoxene concentrate (Pizhemskoe deposit, Russia) and titanium dioxide nanotubes based on it). The materials obtained on the basis of relatively inexpensive and affordable ilmenite-leucoxene ore have the same efficiency as more expensive commercial products.

Keywords: Anatase · Ilmenite-leucoxene ores · Nanotube · Water treatment

1 Introduction

In recent years, advanced oxidation processes have been proposed as alternative methods for eliminating toxic organic pollutants from aquatic systems. Semiconductor heterogeneous photocatalysis is one of the most promising and effective method. This method is environment friendly, because the reaction products of the oxidation of organic pollutants are carbon dioxide and water. A comparative analysis of the economic efficiency of water purification showed that the photocatalytic method was the cheapest (Duduman et al. 2018, Kotova et al. 2016b).

The treatment of water from phenols, in particular, containing chlorine (2, 4, 6 - trichlorophenol, TCP), is an important public health task because of their estrogenic, mutagenic or carcinogenic effects. Their toxicity depends on the degree of chlorination and the position of chlorine atoms in relation to the hydroxyl group. Removing these compounds from the water is necessary to protect both human health and the environment. To produce semiconductor photocatalyst based on titanium dioxide, multi-stage synthesis methods are most often used, using orthotitanium acid or titanium tetrachloride as precursors.

The aim of the work is the comparative evaluation of efficiency of commercially available titanium dioxide (Degussa P25, Anatase Sigma Aldrich), natural (leucoxene concentrate, Pizhemskoe deposit, Russia) and titanium dioxide nanotubes based on natural leucoxene as active photocatalysts (TiNT).

2 Methods and Approaches

Titanium dioxide. Degussa P25 (80% anatase, 20% rutile; Sigma Aldrich, France) was used as photocatalyst without any purification. It has a BET surface (average) of 50 m^2/g, a particle size of 20–50 nm. Anatase (Sigma Aldrich) was used as photocatalyst without any purification. It has a BET surface (average) of 80 m^2/g, elongated particles with a size of 15–30 nm.

Leucoxene concentrate (LC) was obtained from the Pizhemskoe deposit (Russia). Chemical composition (wt%): TiO_2 – 42.12, SiO_2 – 46.57, Fe_2O_3 - 1.04, Al_2O_3 – 7.57, K_2O - 1.61, MnO - 0.06, CaO - 0.13, MgO - 0.37, SO_3 - 0.06, P_2O_5 - 0.17, ZrO_2 - 0.05, NbO - 0.11. The particle size is about 20–40 mcm.

Titanium dioxide nanotubes (TiNT) were obtained using a hydrothermal treat ment procedure. The detailed description is given elsewhere (Kotova et al. 2016a).

The photocatalytic activity of the samples was studied using a test reaction of decomposition of trichlorophenol in Hereaus circular reactor of a volume of 350 cm^3. Vertically to the reactor axis the TQ150 Z2 mercury lamp (150 W, 352–540 nm) was located. The control solutions were analyzed by liquid chromatography (Hypersil C18 reverse phase HPLC column). The solvent was a solution of acetonnitrile in water in a ratio of 3:2. The solvent flow was 0.5 ml/min.

3 Results and Discussion

The initial leucoxen (Fig. 1A) is a mixture of two phases: rutile and quartz. The peaks are clear, which indicates a high crystallinity of these phases. There are weak reflexes of ilmenite and anatase. Leucoxen is a rutile microcrystalline matrix, saturated with the finest inclusions of quartz (Ponaryadov 2017). The synthesized sample (Fig. 1B) is a mixture of two phases: quartz and sodium titanate $Na_2Ti_6O_{13}$. The chemical composition (semi-quantitative): TiO_2 – 74.68%, SiO_2 – 12.64%, Fe_2O_3 – 5.44%, Al_2O_3 – 4.71%, K_2O – 0.93%.

The structural rearrangement at the nanoscale level – formation of titanium di-oxide nanotubes – leads to decreasing band gap: anatase – 3.1, LC –2.8, TiNT – 2.4 eV. Another important parameter is the specific surface area. During formation of titanium dioxide nanotubes we observed increasing specific surface, which is associated with formation of external and internal surfaces. For the studied samples, the specific surface area was: anatase – 80, LC – 13, TiNT – 230 g/m^2.

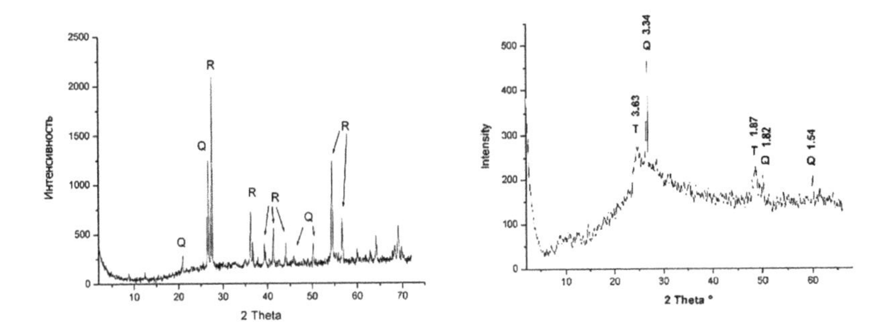

Fig. 1. XRD patterns for leucoxene concentrate, heavy fraction (A) and as-synthesized TiNT (B)

Kinetics of heterogeneous photooxidation reaction in liquid medium in the presence of a catalyst is described by the Langmuir-Hinshelwood model. For the reaction of decomposition of trichlorophenol, the time dependence $\ln(C_0/C)$ is linear, at that the slope ratio gives a constant k_{app}. The time dependence graphs for the studied samples are presented in Fig. 2.

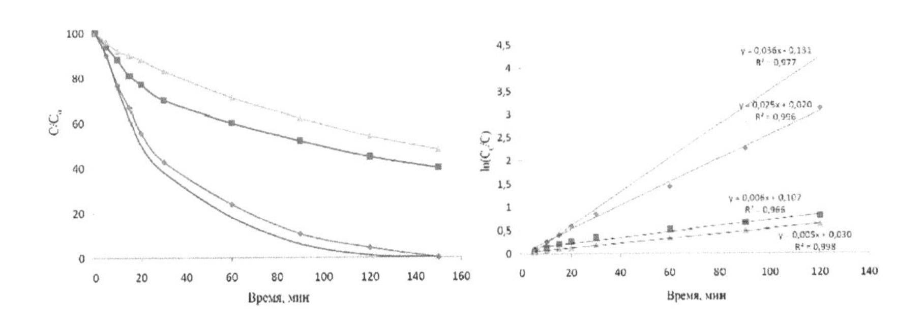

Fig. 2. Curves of decomposition of trichlorophenol in water medium (- leucoxene concentrate, - anatase, - DegussaP25, × - titanium dioxide nanotubes)

The adsorption and decomposition reaction on surface occur simultaneously, most likely, they do not determine the reaction rate. In the initial period of time (0–10 min), trichlorophenol is adsorbed on the sample surface and the reaction rate increases. Upon reaching the full coverage of the surface with adsorbate, the reaction rate is maximal and does not change in the future. Based on the received data, the values k_{app} of reaction constants were calculated: 0.005 for leucoxene concentrate, 0.006 for anatase, 0.025 for Degussa P25, 0.036 for titanium dioxide nanotubes.

Thus, TiNT, produced by the hydrothermal method from ilmenite-leucoxene ore, are competitive photocatalysts in water treatment from organically contaminants in comparison to the above stated synthetic analogues.

4 Conclusions

We studied the dependence of the kinetics of photoinduced decomposition of trichlorophenol in water solutions in the presence of various types of catalysts based on titanium dioxide: commercially available DegussaP25 and anatase (Aldrich), leucoxene concentrate (Pizhemskoe deposit), titanium dioxide nanotubes. We calculated reaction constants of the photoinduced decomposition of trichlorophenol. It is shown that advanced materials on the basis of relatively inexpensive and affordable ilmenite-leucoxene ore have the same efficiency as expensive commercial products.

Acknowledgements. This research was supported by UB RAS project № 15-18-5-44 and project AAAA-A17-117121270037-4 "Scientific basis for effective development and use of the mineral resource base, development and implementation of innovative technologies, geological and economic zoning of the Timan-North Ural region".

References

Duduman CN, de Salazar y Caso de Los Cobos JMG, Harja M, Barrena Pérez MI, Gómez de Castro C, Lutic D, Kotova O, Cretescu I (2018) Preparation and characterisation of nanocomposite materials based on TiO_2-Ag for environmental applications. Environ Eng Manag J 17(4):2813–2821

Kotova O, Ozhogina E, Ponaryadov A, Golubeva I (2016b) Titanium minerals for new materials. In: IOP conference series: materials science and engineering, p 012025. https://doi.org/10.1088/1757-899x/123/1/012025

Kotova OB, Ponaryadov AV, Gömze LA (2016a) Hydrothermal synthesis of TiO2 nanotubes from concentrate of titanium ore Pizhemskoe deposit (Russia). Vestnik IG Komi SC UB RAS 1:34–36

Ponaryadov AV (2017) Mineralogical and technological features of ilmenite-leucoxene ores of Pizhemskoe deposit, Middle Timan. Vestn Inst Geol Komi SC UB RAS 1:29–36. https://doi.org/10.19110/2221-1381-2017-1-29-36 (in Russian)

5

Kinetic Features of Formation of Supramolecular Matrices on the Basis of Silica Monodisperse Spherical Particles

D. Kamashev(✉)

Institute of Geology Komi SC UB RAS, Syktyvkar, Russia
kamashev@geo.komisc.ru

Abstract. We have established that under such conditions, when the formation of a supramolecular structure from spherical silica particles 220–320 nm in diameter is limited by the rate of introduction of particles into the sedimentation zone (sedimentation deposition in a constant cross section tube), the particle deposition rate, as well as the formation rate of the supramolecular structure, is strictly linear. At the same time, under conditions of excess of disperse phase in the zone of formation of the supramolecular structure (sedimentation deposition in a tube with a modified cross section), the deposition rate is also linear, but there is some delay in formation of the supramolecular structure in time, the larger the smaller is the particle size of the disperse phase. After the formation of the supramolecular structure is completed, a region with an increased concentration of the disperse phase remains, the height of which is greater, the smaller is the particle size. It is shown that a certain concentration of the disperse phase is necessary to begin the formation of a supramolecular structure, below which the formation of a supramolecularly ordered structure does not occur. The concentration of the disperse phase, necessary for beginning of formation of a supramolecular structure, is a constant value that does not vary with time and depends only on the size of the particles.

Keywords: Supramolecular structure · Monodisperse spherical silica particles

1 Introduction

The supramolecularly ordered structures, based on monodisperse spherical silica particles, generated interest relatively long ago, back in the 70s of the last century, in connection with attempts to synthesize artificial analogs of noble opal on their basis (Stober et al. 1968). The overwhelming majority of studies of that time were aimed at developing conditions for synthesis of spherical silica particles and supramolecular structures based on them, and ended with the development of methodological bases for production of synthetic noble opals (Deniskina et al. 1987). However at present the supramolecular silica particles are more often considered as promising objects for synthesis of new composition materials, photonic crystals and nanostructured materials based on them. The 3D ordered closest packing of monodisperse silica spheres is an ideal matrix for creating a wide class of new nanostructured materials, but it greatly increases requirements for the monodispersity of particles, their size and the defects of derived supramolecular structures (Kamashev 2012). With this aim and within the task to develop the basis of synthesis of supramolecularly ordered matrices we carried out experimental studies of the rate of precipitation of silica particles and formation of a supramolecular structure on their basis in various conditions (growth modes).

2 Methods and Approaches

Monodisperse spherical silica particles with radius 109, 138 and 158 nm were synthesized by Stober-Fink method (Stober et al. 1968). The sizes of the obtained particles were determined by Photocor Complex dynamic light scattering spectrometer at a laser wavelength 661 nm, a scattering angle 90°, and a correlation function accumulation time 20 min. The initial concentration of the dispersed phase particles was about 2 wt.%. The deposition rate of the silica particles was measured in glass tubes 750 mm high and 20 mm in diameter, with the control of deposition boundary advancing every 30 days with accuracy 0.5 mm. The average daily temperature was also taken into account to calculate the temperature correction associated with the expansion of the dispersion environment. The following values were obtained for the silica particles of different radius (Fig. 1).

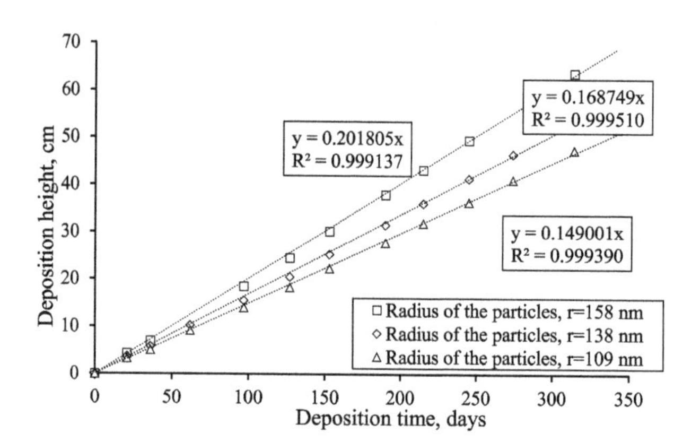

Fig. 1. Dependence of height of deposition of silica particles of different sizes over time. The coefficient before x (0.201805, 0.168749 and 0.149001) in the equation of approximating curve represents the rate of particle deposition, cm/day.

To measure the rate of formation of the supramolecular structure, the deposition of particles was carried out in glass tubes with a narrowing in the lower part, which reduced the area (concentration of suspension) by about 80 times. This mode of formation of the supramolecular structure is characterized by a constant increased content of the dispersed phase, unlike deposition in tubes with a constant cross section, where the formation of region with a high content of particles is limited by their deposition rate. The obtained data on the rate of formation of a supramolecularly ordered structure are presented in Fig. 2. At such rates one monolayer of silica particles with radius 158 nm is formed approximately within 90 s.

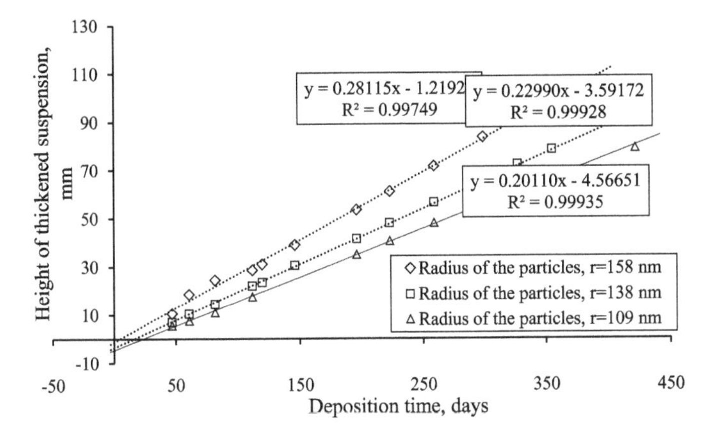

Fig. 2. Dependence of height of a supramolecular structure formed from silica particles of different size over time. The coefficient before "x" (0.28115, 0.22990 and 0.20110) in the equation of approximating curve is a growth rate of the supramolecular structure in conditions of high concentration (excess) of the dispersed phase, mm/day.

3 Results and Discussion

Our data testify to that the particle deposition rate (Fig. 1) is linear and starts at the origin and strictly obeys the Stokes equation throughout the time. In turn, the analysis of data on the formation rate of the supramolecular structure (Fig. 2) shows that the formation rate of the ordered structure is also linear, however, the process of supramolecular crystallization does not begin immediately, but after some time, due to creation of "supersaturation" or some increased concentration of particles of dispersed phase in the bottom region. This time depends on size of particles and increases with their decrease.

As a result of deposition of monodisperse silica particles and formation of a supramolecular structure we noted that initially a zone with an increased concentration of dispersed phase was formed in the bottom region. The boundaries of this zone were clearly expressed, it possessed a constant height, both in the process of the beginning of deposition of particles, and after its completion. The height of the condensed

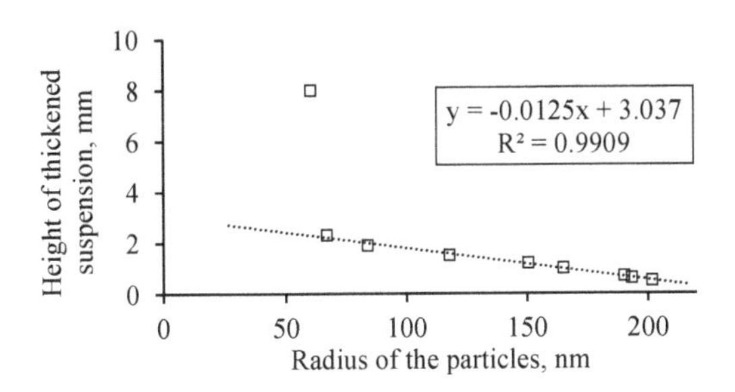

Fig. 3. Dependence of thickened suspension height on the particle size of the disperse phase after completion of deposition.

suspension zone after deposition of particles was constant and depended only on size of dispersed particles (Fig. 3) with the exception of particles of radius less than 60 nm. In this case a similar zone with clearly defined boundaries was not formed, and the supramolecular structure was not formed.

4 Conclusions

The obtained experimental data on deposition rate of monodisperse spherical silica particles with formation of a supramolecular structure showed that its formation was the result of a kind of second-order phase transition (self-organization of particles). This transition was associated with a threshold concentration of particles of dispersed phase in the area of deposition, below which the formation of the supramolecular structure did not occur. When this value was exceeded, the process of formation of a supramolecular structure started, and the rate of its formation was strictly linear and independent of the concentration of the dispersed phase. If for some reason the threshold concentration of particles was not reached (for example, very small sizes, low density of particles or high viscosity of the dispersion medium), then the formation of a supramolecular structure did not occur.

Acknowledgements. The work was accomplished with partial financial support by UB RAS program No. 15-18-5-44, and RFBR No. 19-05-00460a.

References

Deniskina ND, Kalinin DV, Kazantseva LK (1987) Noble opals. Novosibirsk, Nauka, Sib. otd. (Trudy institute geologii I geofiziki) 693, p 184

Kamashev DV (2012) Synthesis, features and model of formation of supramolecular silica structures. Phys Chem Glass 38(3):69–80

Stober W, Fink A, Bohn E (1968) Controlled growth of monodisperse silica spheres in the micron size range. J Colloid Interface Sci 26:62–69

6

Rational Usage of Amorphous Varieties of Silicon Dioxide in Dry Mixtures of Glass with Specific Light Transmittance

N. Min'ko(✉) and O. Dobrinskaya

Belgorod State Technological University named after V G Shukhov, Belgorod, Russia
minjko_n_i@mail.ru

Abstract. The paper studied high-silicon amorphous rocks from the perspective of their application for glass production of different purpose. The results contained data of calculation of dry mixtures for producing heat-protective glass using amorphous varieties of silicon dioxide. The obtained glass specimens were melted and studied for spectral characteristics.

Keywords: Amorphous silicon dioxide · Light transmittance · Dry mixture · Heat-protection glass

1 Introduction

High-silicon amorphous rocks as raw materials have a wide range of valuable features. Primarily, this is the amorphous (metastable) state of silicon dioxide (Kondrashov and Kondrashov 2013). Moreover, one can distinguish some peculiarities of amorphous varieties that can be regarded as drawbacks (Manevich et al. 2012):

- instability of chemical composition;
- SiO_2 is accompanied by other components (up to 40%) that can play the role of auxiliary raw materials;
- increased content of aluminum oxides and iron.

One of the main glass spectral characteristics is light transmittance. The main components that affect light transmittance of glass products and that should be strictly controlled are oxides of coloring metals that are encountered in conventional raw materials (dolomite, feldspar concentrate, sands). These compounds include iron oxides; their content in glass is strictly regulated:

- sheet glass – 0.09–0.12%;
- heat-protective – 0.6–0.7%;
- clear container glass – 0.1 ± 0.01%;
- brown container glass – 0.8 ± 0.1%;
- green container glass – not regulated.

The composition of amorphous varieties of silicon dioxide (Distantov 1976) is characterized by increased content of iron oxides (Table 1), which impedes their wide application in production of glass articles.

Table 1. Variation of chemical composition of amorphous silicic raw materials

Rock	Content of oxides [wt.%]							
	SiO_2	Al_2O_3	Fe_2O_3	CaO	MgO	K_2O	Na_2O	TiO_2
Diatomites	73.0–90.0	3.3–7.5	2.0–5.2	less than 0.6	0.6–1.7	less than 1.0	less than 0.5	less than 0.3
Opokas	52.1–91.4	2.5–15.4	1.0–5.0	0.43–17.1	less than 2.48	0.6–4.0	0.1–1.0	less than 0.2
Pearlites	68.5–75.3	11.2–16.3	less than 3.0	0.5–2.0	less than 1.0	1.5–4.0	2.0–6.2	0.1–0.5
Tripolites	35.3–86.7	2.5–11.6	0.3–3.4	0.4–31.2	0.2–1.6	0.85–2.1	less than 0.5	less than 0.2

Taking into account increased content of iron and aluminum oxides, these rocks can be used as aluminum- or iron-containing raw materials in production of heat-protection glass or dark-glass containers.

2 Methods and Approaches

Current work assesses amorphous silicon dioxide (ASD) as aluminum-containing raw material that can partially replace quartz sand and other conventional raw materials (Table 2).

Table 2. Composition of average samples of amorphous silicon dioxide varieties

Rock	Content of oxides [wt.%]							
	SiO_2	Al_2O_3	Fe_2O_3	CaO	MgO	K_2O	Na_2O	TiO_2
Diatomite	87.22	6.79	2.22	0.431	1.25	1.08	0.245	0.265
Opoka	92.47	3.38	1.32	0.506	0.712	0.88	–	0.174
Pearlite	72.0	16.45	1.06	0.863	0.422	4.30	4.21	0.145
Tripolite	79.93	10.99	3.17	0.838	1.81	1.99	0.273	0.687

To conduct experimental studies, the dry mixtures for heat protective glass with ASD were calculated. The glass was melted from the dry mixtures using pearlite. The content of iron oxides in pearlite is insufficient for production of heat-protective glass, which necessitates the introduction of iron containing material such as magnetite.

The glass was melted in an electric kiln with silicon carbide heating elements at maximum melting temperature of 1420 °C. Another batch of glass was melted without pearlite.

Spectral light transmittance in the visible range was measured automatically on SF-56 spectrophotometer (Russia). The specimens were prepared by mechanical grinding and polishing on laboratory setups.

3 Results

The results showed that the application of ASD for preparation of dry mixtures allowed reducing number of conventional materials for glass melting (Table 3).

Table 3. Economy of conventional raw materials after replacement by ASD, %

Rock	Raw material			
	Sand	Soda	Creta	Dolomite
Opoka	**72.7**	2.44	1.61	19.71
Pearlite	43.8	**20.15**	**38.7**	**40.38**
Diatomite	41.7	6.5	4.08	25.14
Tripolite	27.6	7.79	7.66	27.06

The glass produced in the laboratory differed in color (Fig. 1), which depends on the shift of equilibrium $Fe^{2+} \leftrightarrow Fe^{3+}$.

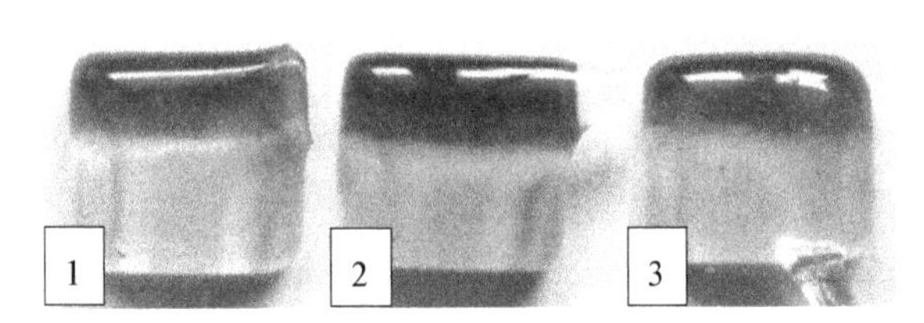

Fig. 1. Specimens of sheet glass: 1 – with pearlite; 2 – with pearlite and magnetite; 3 – with pearlite and magnetite (+ coal)

We studied spectral characteristics of the glass specimens: sheet glass melted from conventional components; sheet glass with addition of pearlite as aluminum-containing raw material; sheet glass with addition of magnetite (magnetite was reduced by coal) (Fig. 2).

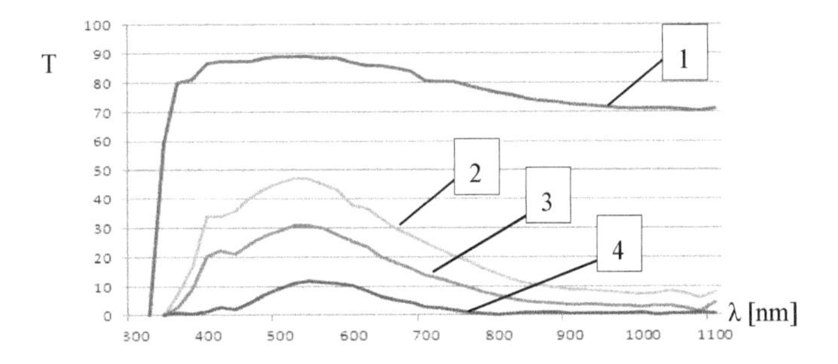

Fig. 2. Spectral light transmittance of sheet glass: 1 – conventional raw materials; 2 – with pearlite; 3 – with pearlite and magnetite; 4 – with pearlite and magnetite (+ coal)

The data from Table 4 demonstrated that in the visible spectrum, the light transmittance of the glass with application of pearlite differed from glass melted with conventional raw materials. The highest light transmittance was shown by sheet glass melted from conventional raw materials, since the content of iron in quartz sand was minimal (0.022 wt%).

Table 4. Light transmittance of sheet glass specimens calculated for the thickness of 10 mm

Specimen	Light transmittance T [%]				Diathermancy index = 10^{-1} T_{1100}
	λ = 570 nm		λ = 1100 nm		
	As per equation	As per nomogram*	As per equation	As per nomogram*	
1	88.6	88.1	67.4	67.7	6.7
2	48.4	48.8	8.4	8.2	0.8
3	30.8	30.5	5.1	4.8	0.4
4	12.3	12.0	3.7	3.2	0.3

* - Amosov's nomogram

The heat-protective characteristics are assessed by light transmittance at λ = 1100 nm. In the infrared range of the spectrum, the light transmittance of specimens 2, 3 and 4 reduces due to the presence of impurities of Fe^{2+}, which provisions the heat-protective characteristics.

In IR-range, the light transmittance of studied glasses is different (from 3 to 67%), i.e. high heat-protective properties are possessed by specimens 3 and 4, the IR-light transmittance is 3–5%. However, the production of glass with such content of FeO is unreasonable due to low light transmittance in the visible spectrum. Such dry mixtures can be used for special glass.

4 Conclusions

The calculations allowed determining the amorphous varieties of silicon dioxide that could replace most conventional materials in dry mixtures. They were: opoka (as a replacement for quartz sand) for both aluminum- and iron-containing raw materials, and pearlite capable of replacing the biggest amount of soda, creta and dolomite when using it as an iron-containing material.

Additional studies of glass properties produced from dry mixtures with pearlite were required (spectral characteristics, crystallization capacity and impact of redox conditions of melting). Nevertheless, the work showed that pearlite could be used in the technology of heat-protective glass.

Acknowledgments. The work is realized in the framework of the Program of flagship university development on the base of Belgorod State Technological University named after V.G. Shukhov, using the equipment of High Technology Center at BSTU named after V.G. Shukhov.

References

Distantov AG (1976) Silicon rocks of USSR. Tatar Press, Kazan
Kondrashov VI, Kondrashov DV (2013) Perspectives of synthesis of industrial compositions of float-glass on the basis of crystalline and amorphous silicon dioxide. GlassRussia. 3:31–33
Manevich VE, Subbotin RK, Nikiforov EA, Senik NA, Meshkov AV (2012) Diatomite as silica-containing material for glass industry. Glass Ceram 5:34–39

7

Development of Technology for Anti-Corrosion Glass Enamel Coatings for Oil Pipelines

E. Yatsenko(✉), A. Ryabova, and L. Klimova

South Russian State Polytechnical University (NPI), Novocherkassk, Russia
e_yatsenko@mail.ru

Abstract. Among anticorrosive coatings for steel products, glass-enamel glass is the most reliable and versatile, based on aluminoborosilicate glasses of the SiO_2-Al_2O_3-B_2O_3-R_2O system. This anti-corrosion coating allows to increase the chemical resistance of the internal surface of pipelines to various groups of reagents. Therefore, in the course of the study, a previously developed composition was modified by introducing oxides SrO, ZrO_2, CaO, MoO_3, Li_2O and their acid and alkalinity of enamel frits and coatings based on them and it was found that the addition of strontium and zirconium oxide in the amount of 2% was optimal.

Keywords: Oil pipelines · Anti-corrosion coatings · Glass-enamel coatings · Steel protection · Chemical resistance · Acid and alkali resistance

1 Introduction

Currently, the oil industry of the Russian Federation is developing rapidly and efficiently. However, corrosion of equipment and facilities in the oil and gas industry is one of the main reasons for reducing their performance, causing huge economic losses and environmental damage. Presently, protective coatings (bitumen, epoxy, polyurethane, etc.) have become widely used to protect oil pipelines, among which glass-enamel coatings with high chemical resistance and thermomechanical properties, in particular, heat resistance, reaching 500 C, occupy a special place. Due to the fact that the silicate-enamel coating for steel pipelines is exposed to an aggressive environment containing hydrocarbons and formation water, in which chlorides, sulfates and organic acids are present, as well as up to 10% hydrogen sulfide and 10% carbon dioxide. The coating composition was based on an acid-resistant glass composition (Ryabova et al. 2018) with a high content of quartz and low boric anhydride and alkaline oxides, which will improve the chemical resistance of enamels and extend the range of their roasting (Yatsenko et al. 2018).

Also, when choosing the type of protective coating, the following factors should be taken into account: operating conditions, composition of the transported medium, temperature and pressure in the system, speed and nature of the flow movement, presence of abrasive particles in the fluid flow, composition and properties of associated petroleum gas (APG), presence of asphalt-resin-paraffin deposits (AFS), the manifestation of the life of microorganisms.

Therefore, the purpose of these studies was the development of anticorrosive glass-enamel coatings to protect steel pipelines from medium carbon steel 32G2S and the study of factors affecting the process of their defect-free formation and technical and operational properties.

2 Methods and Approaches

The technology of enameling the internal surface of pipelines includes the following technological stages: preliminary annealing of steel pipes at a temperature of ~750 °C in order to remove contaminants and decarburize the outer layers of steel; mechanical preparation of the inner surface of pipes using shot blasting using single or multiple shot blasting units to remove scale and roughen for better adhesion to the glass-enamel coating; preparation of enamel slip suspension on the basis of finely ground glass granulate; application of slip suspension on the inner surface of a horizontal pipe by sprinkling using an impeller; drying at a temperature of 100–120 °C and induction roasting at a temperature of 860–880 °C.

The aim of the work was to develop the composition and technology of applying a glass-enamel single-layer coating for medium-carbon steels, which has high rates of chemical and abrasive resistance and is capable of forming a defect-free smooth coating on the steel surface. To solve this problem, the aluminoborosilicate system SiO_2-Al_2O_3-B_2O_3-R_2O was chosen as the most acceptable in the single-layer enamelling technology, which was modified by introducing compounds such as SrO, ZrO_2, CaO, MoO_3, Li_2O, in order to improve chemical resistance and defect-free formation in the form of through pores. The introduction of enamels strontium, calcium and zirconium oxides that are insoluble in oxides helps to reduce the leaching of alkali and alkaline earth cations when exposed to acid coating. Lithium oxide together introduced with oxides of sodium and potassium contributes to the chemical resistance of enamel coatings due to the manifestation of polyalkalnochnogo and polycationic effects. The amount of additives introduced into the charge was 2 wt.%, Since the introduction of additives less than 1% slightly affects the properties of enamel coatings, and more than 2% can greatly affect the change in the technological properties of the melt.

Next, the compounded mixtures were boiled at a temperature of 1350 °C for 1 h in an electric oven in alundum crucibles, subjected to wet granulation and applied in the form of finely ground enamel slip to samples of 32G2S steel. After drying and firing at a temperature of 850 °C, the resulting enamel coatings were tested to study the effect of modifying oxides on the structure and properties of enamels.

To assess the corrosion resistance, tests were carried out to determine the acid resistance characterized by weight loss after exposure to 20% boiling hydrochloric acid and alkali resistance - weight loss after exposure to 8% sodium hydroxide solution. For frits, the tests were carried out by the grain method, and for enamel coatings, the impact on their surface.

3 Results and Discussion

For the developed modified frits and enamel coatings, chemical resistance to various groups of reagents for weight loss after boiling in acid and alkaline solutions was studied. The test results are presented in Table 1.

Table 1. Indicators of the properties of modified frits and enamel coatings depending on the composition

Name of enamel	Modifying additive	Chemical resistance			
		Frites, %		Coatings, mg/cm^2	
		20%-HCl	8%-NaOH	20%-HCl	8%-NaOH
10-0	Without additives	0,38	0,83	0,22	0,78
10-1	SrO	0,32	0,99	0,16	0,45
10-2	ZrO_2	0,40	1,20	0,20	0,60
10-3	CaO	0,38	0,75	0,21	0,63
10-4	MoO_3	0,54	0,98	0,22	0,68
10-5	Li_2O	0,42	0,87	0,23	0,63

The results obtained allow us to conclude that the introduction of modifying additives into the glass matrix composition has a significant impact on anti-corrosion properties.

For all compositions, the mass loss of frits after boiling is quite significant, since the specific surface of glass powders is much larger than the surface of the burned enamel coating. However, the composition of frits modified by strontium and zirconium oxides is characterized by less mass loss, which is caused by their positive effect on the increase in the packing density of the structural amorphous network, due to the large radius of these ions. Molybdenum oxide has almost no effect on chemical resistance, but it has a positive effect on the continuity of the coating contributing to a smoother formation and the absence of coating defects, due to a decrease in the surface tension of the enamel melt. Calcium oxide in such an amount does not affect the chemical resistance of frits and coatings. Lithium oxide increases chemical resistance and frit and coatings, although it is alkaline in itself, but it has an inhibitory effect due to the presence of several alkali cations (Na_2O and K_2O).

4 Conclusions

As a result of the research, the composition and technology of producing anticorrosive glass-enamel coatings for the internal protection of steel pipelines based on the SiO_2-Al_2O_3-B_2O_3-R_2O aluminoborosilicate system has been developed. The effect of various modifiers of the vitreous matrix on the acid and alkali resistance was studied, and it was found that the strontium and zirconium oxides in an amount of 2% are optimal.

Acknowledgements. The work was done with the financial support of the Russian Science Foundation under the agreement No. 18-19-00455 “Development of technology for the integrated protection of pipelines for oil and gas operated in the Far East of Russia” (headed by Yatsenko E. A.).

References

Ryabova AV, Yatsenko EA, Klimova LV, Goltsman BM, Fanda AY (2018) Protection of steel pipelines with glass-enamel coatings based on silica-containing raw materials of the far east of Russia. Int. J. Mech. Eng. Technol. 9(10):769–774

Yatsenko EA et al (2018) Investigation of chemical processes that ensure the adhesion strength of glass-enamel coating with steel pipelines. Butlerov Commun 56(11): 122–127

8

The use of Karelia's High-Mg Rocks for the Production of Building Materials, Ceramics and other Materials with Improved Properties

V. Ilyina(✉)

Institute of Geology KarRC RAS, Petrozavodsk, Russia
Ilyina@igkrc.ru

Abstract. The possible use of high-Mg host rocks, such as serpentinite and pyroxenite, from the Aganozero chromium ore deposit and serpentinite from the Ozerki soapstone deposit, Republic of Karelia, for the production of heat-insulating building materials, ceramic pigments and filters for purification of technogenous solutions is assessed. The results of analysis of the mineralogical compositions of serpentinites and pyroxenite, as well as the physico-mechanical properties (strength, heat conductivity, shrinkage upon roasting, moisture resistance, etc.) and structural characteristics of the ceramic and heat-insulating materials produced on their basis are reported.

Keywords: Serpentinite · Pyroxenite · Ceramics · Heat conductivity · Mechanical strength · Facing material

1 Introduction

Non-conventional high-Mg rocks can be used for the production of ceramic, building and other materials because they are widespread but mainly because of the chemical, mineral and structural characteristics of their mineral constituents: periclase, forsterite, diopside, augite, enstatite and serpentine. The main high-Mg mineral constituents of ceramics suffer phase transformations upon heating, as a result of the disintegration and recrystallization of their lattice (Deere et al. 1965), forming crystalline phases that improve the physic-mechanical properties of ceramics. As forsteritic ceramics suffers no polymorphic transformations, it does not age and is mechanically strong. It is used for the production of dielectrics, heat-insulating materials, facing ceramics and filters for water purification. The aim of the present project is to study serpentinites, kemistites (rocks of hydrotalcite-serpentine composition) and pyroxenites that host chromium ores at the Aganozero deposit and serpentinites at the Ozerki soapstone deposit and to use them for the production of ceramics with a dominant forsteritic crystalline phase, filters for purification of technogenic solutions and ceramic pigments.

2 Methods and Approaches

The mineral composition of the analyzed samples was studied in the IG KarRC RAS by optical microscopy methods, X-ray phase analysis (XPA) and thermal analysis (TA). Rock-forming minerals were studied by Vega II LSH scanning electron microscope with INCA Energy 350 energy dispersion analyzer. X-ray phase analysis was performed by ARL X'TRA diffractometer with CuKl radiation. The physico-mechanical properties of the materials and ceramics were assessed in accordance with State All-Russia standards.

3 Results and Discussion

Differences in the mineral composition of serpentinites affect their chemical composition. Aganozero kemistites and serpentinites are the richest in magnesium (36–38%) and contain minor quantities of impurities (0.1–0.5% Al_2O_3 and 0.24–0.5% CaO) and elevated quantities of crystallization water (loss on ignition is 15–18.5%). Ozerki serpentinite is rich in MgO (36.92%) and contains Al_2O_3 (2.2) and CaO (0.22%) as impurities. All the samples are iron-rich (3.46–10.02% Fe_2O_3, 1.72–3.9% FeO). The mineralogo-analytical study (RPA, TA) of serpentinites showed that Ozerki serpentinites consisted of 89.1% fine-grained lamellar antigorite aggregate and that Aganozero serpentinites contained 78% lizardite. Ore minerals are represented by magnetite and ilmenite. The main minerals of natural pyroxenite (wt%) are augite (67, 2), forsterite (4, 3), enstatite (23, 7) and serpentine (4, 8). Kemistite-based porous heatinsulating ceramics was developed and its properties were studied (Patent no. 2497774, 2013). Electron microscopy study has shown that forsterite, produced by serpentine recrystallization, is the main crystal-line phase of heat-insulating ceramics. The properties of heat-insulating ceramics are shown in Fig. 1.

Pyroxenite (20–70%) and hydromica-based facing ceramics (Il ina et al. 2017) displays water absorption of 13–15.8% at a roasting temperature of 900–1100 °C, which is consistent with the current standards. At 1200 °C, the water absorption of all the masses decreases rapidly from 0 to 1.2%. Their bending strength is 9 10 MPa.

Pyroxenite-based ceramic pigment has been developed. Unlike the well-known pigment, it can be used to obtain a stable color after roasting over a wide temperature of 750–1250 °C.

The results obtained (Ilyina et al. 2018) show that Ozerki serpentinite can be used for the production of an Mg-silicate reagent for the removal of heavy metals from solutions, e.g. the removal of heavy metal compounds from highly polluted technogenous solutions by filtration through loading from a granulated reagent.

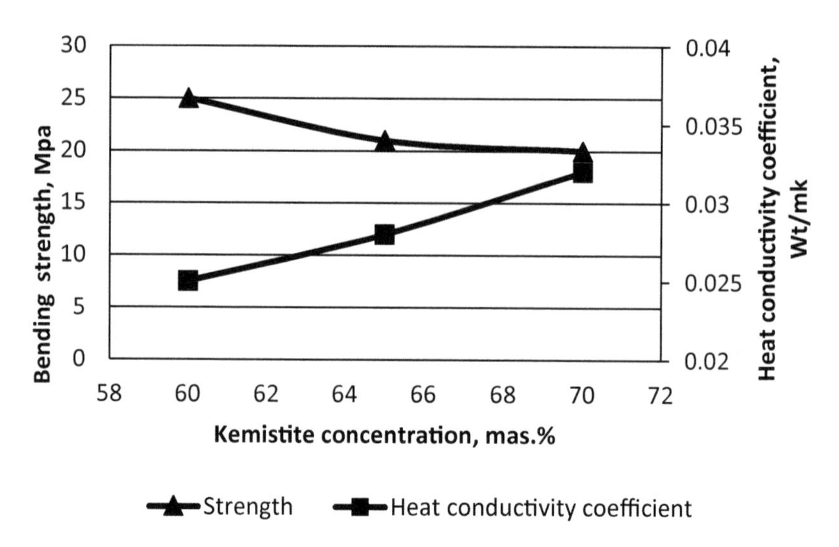

Fig. 1. Dependence of variations in the properties of ceramics on kemistite concentration

4 Conclusions

The materials developed display a porous structure and, consequently, low density, low heat conductivity and high strength. Hence, they can be widely used in industrial and civil engineering for the insulation of buildings, for the heat insulation of the hot surfaces of equipment (furnaces and pipelines), for intensifying high temperature processes and for fuel saving.

References

Deere WA, Haui RA, Zoosman J (1965) Rock-forming minerals. Chain silicated. vol 2, p 405

Il'ina VP, Inina IS, Frolov PV (2017) Ceramic mix based on pyroxenite and low-melting clay. Glass and Ceramics pp 1–4

Ilyina VP, Kremenetskaya IP, Gurevich BI, Klimovskaya EE, Ivashevskaya SN (2018) The study of serpentinized ultramafics from the Kareli-an-Kola Region and the production of a Mg-rich-silicate reagent on their basis for the removal of heavy metals from solutions. In: 18th International Multidisciplinary Scientific GeoConference SGEM2018: Conference proceedings. STEF 92 Technology Ltd., 51 "Alexander Malinov" Blvd., 1712 Sofia, Bulgaria, Energy and Clean Technologies Issue: 4, 2. 2 July – 8 July 2018. Albena, Bulgaria, vol 18 (13), pp 207–213

Ilyina VP, Shchiptsov VV, Frolov PV (2013) Raw mixture for the pro-duction of a porous heat-insulating material, Bull no 31 Patent no 2497774 RF, MPC SO4B 33/132

9

Heating Rate and Liquid Glass Content Influence on Cement Brick Dehydration

V. Strokova(✉) and D. Bondarenko

Department of Material Science and Material Technology, Belgorod State Technological University named after V.G. Shukhov, Belgorod, Russia
vvstrokova@gmail.com

Abstract. Peculiarities of Portland cement dehydration in different hydrate phases with sodium water glass have been given. Three endoeffects were determined during non-isothermal heating, connected with ettringite dehydration and water extraction at temperature ranges 98.7–110.0 °C, calcium hydroxide decomposition at temperature ranges 439.4–450.7 °C and secondary carbonates decomposition at temperature ranges 657.4–669.3 °C. We experimentally proved that the rates of dehydration of hydrated Portland cement was significantly influenced by the liquid glass concentration. Optimum liquid glass content was grounded in the protective layer of composite finishing material, modified with low-temperature plasma.

Keywords: Plasma-chemical modification · Dehydration · Cement brick · Soda water glass

1 Introduction

Plasma-chemical modification is one of the promising technologies of creating protective–decorative coatings in the manufacture of finishing construction materials for building and construction faces (Bondarenko et al. 2018a, b; Bessmertny et al. 2018; Bondarenko et al. 2016; Volokitin et al. 2016). Dehydration, melt formation and accumulation during plasma melting take second fractions, and the surface is heated up to 2000 °C. As a result of high temperature impact hydro silicate dehydration in the surface layers can result in micro cracking and protective-decorative coating softening, as well as coating adhesion strength reduction and lowering of cold resisting properties.

Insufficient technology elaboration on reducing heat impact consequences and dehydration minimizing plasma melting of cement concrete does not allow wide application of these materials on the national market. That is why the main task in developing treatment technologies for materials based on cement matrix is composition development for protective coating which eliminate these processes.

2 Methods and Approaches

To prove the efficiency of Portland cement and liquid glass application in the protective coating during manufacture of composite finishing material with plasma surface treatment the samples were prepared at water\concrete ratio 0.3 from pure Portland cement (CEM I 42,5 H) and with 5 and 10% of soda water glass (ρ = 1,4 g/sm^3, silica modulus 2.8) of water of mixing. After hardening at normal conditions during 28 days, the samples were exposed to differential-thermal analysis.

Plasma-chemical surface modification is done in non-isothermal conditions at heating rate 3000 °C/min. It makes impossible to study dehydration in real conditions of plasma heating. This process was simulated with simultaneous thermal analysis device Netzsch STA 449 F3 Jupiter at heating rates 5 °C/min and 10 °C/min with maximum heating rate 1000 °C.

3 Results and Discussion

The thermograph of pure hydrated Portland cement shows three endoeffects (Table 1). The first endoeffects, in the temperature range 98.7–110.0 °C in the low temperature area, is connected with ettringite dehydration ($Ca_6Al_2(SO_4)_3(OH)_{12}·26H_2O$) and water extraction. Endoeffects of these two processes superimpose each other. The second is connected with calcium hydroxide dehydration ($Ca(OH)_2$) and happens at temperature ranges 439.4–450.7 °C. The third endoeffects (657.4–669.3 °C) is connected with the secondary hydro carbonates dehydration ($CaCO_3$). Complete water extraction is at 900 °C.

Table 1. Changing of endoeffects with the introduction of liquid glass and heating range 5 and 10 °C/min

Endoeffect producer	Pure hydrated Portland cement		Portland cement with 5% liquid glass		Portland cement with 10% liquid glass	
	Heating range, °C/min					
	5	10	5	10	5	10
Ettringite and physically-coupled water	98.7	110.0	92.6	106.4	92.8	108.0
Calcium hydroxide	439.4	450.7	437.8	450.6	438.0	452.6
Secondary carbonate and hydro silicate	657.4	669.3	662.0	683.1	663.0	693.6

Similar results were received with the hydrated Portland cement after adding 5 and 10% liquid glass (Table 1).

A positive effect of liquid glass adding on secondary carbonate and hydro silicate endoeffects, which are responsible for cement brick softening and micro cracking at higher temperature range, can be explained by effect of encapsulation of hydrate phases with coating of liquid glass.

Adding sodium silicate solute into Portland cement in the amount 5 and 10% takes down mass loss (TG) in ettringite dehydration area (Fig. 1). But in high temperature area dehydration intensity increases up to 2–3%, it is especially notable with 10% liquid glass.

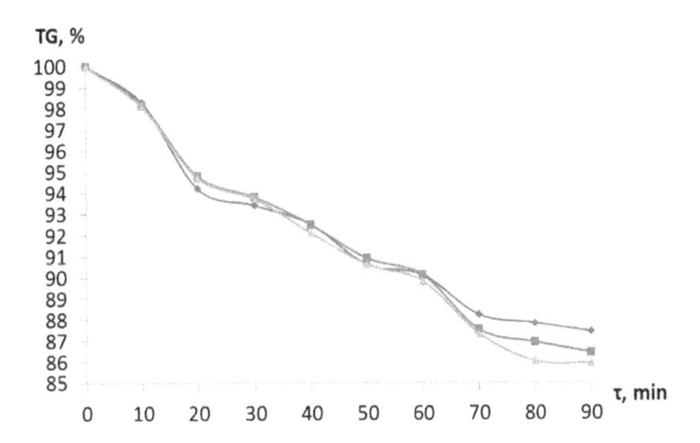

Fig. 1. Dependence of water loss on time at heating range 10 °C/min: — Portland cement; — Portland cement with 5% liquid glass; — Portland cement with 10% liquid glass

The highest dehydration rate is in low temperature area (Fig. 2), which is caused by ettringite dehydration (first climax). The second and third climaxes are connected with dehydration of calcium hydroxide, secondary carbonate and different hydro silicates, are below the first climax by magnitude.

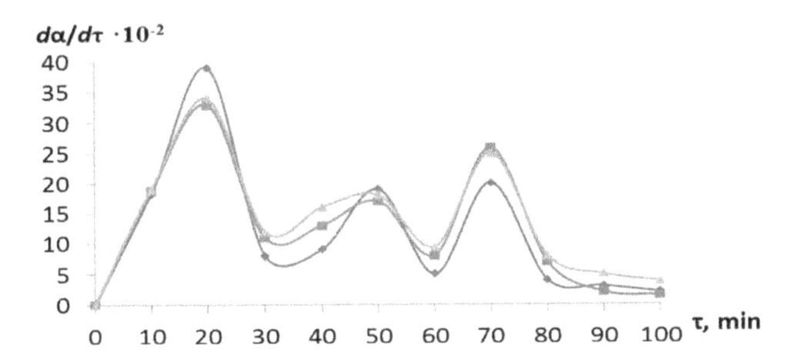

Fig. 2. Dependence of water loss on time at heating rates 10 °C/min: — Portland cement; — Portland cement with 5% liquid glass; — Portland cement with 10% liquid glass

Dehydration rate decrease during the first and the second climax for cement brick with 5 and 10% of liquid glass has a significant impact on micro cracking minimization in the surface layer of protective–decorative coating of composite finishing material at plasma-chemical modification. Dehydration intensification can result in microcracks increase and reduce to zero the positive influence of liquid glass in the coating. This effect takes place at adding 10% liquid glass into Portland cement. Thus, the received laws analysis of mass loss and dehydration rate of the studied compositions drew to a conclusion that the optimal is liquid glass component in the ratio 5% of mixing water.

4 Conclusions

The influence pattern character has been determined of liquid glass content on ettringite dehydration rate, calcium hydroxide and secondary carbonate, which is in endoeffects shift in low temperature area to lower temperatures, and in high temperature area to the area of high temperatures. This minimizes dehydration in low temperature area, cement brick softening and micro cracking and as a result provides adhesion strength improvement of protective-decorative coating with the concrete layer.

Acknowledgements. The work is realized with the financial support of the Grant of the President for scientific schools, No. NSh-2724.2018.8, using equipment of High Technology Center at BSTU named after V.G. Shukhov.

References

Bessmertny VS, Puchka OV, Bondarenko DO, Antropova IA, Bragina LL (2018) Plasmochemical modification of wall building materials. Constr Mater Prod 1(2):11–18

Bondarenko DO, Bondarenko NI, Bessmertnyi VS, Strokova VV (2018a) Plasma-chemical modification of concrete. Adv Eng Res 157:105–110

Bondarenko NI, Bessmertnyi VS, Borisov IN, Tymoshenko TI, Burshina NA (2016a) The concrete with protective and decorative coverings on the basis of alyuminatny cements which are melted off by the plasma stream. Bulletin of Belgorod State Technological University named after V.G. Shukhov, vol 2, pp 6–12

Bondarenko NI, Bondarenko DO, Burlakov NM, Bragina LL (2018b) Investigation of influence of plasmochemical modification on macro- and microstructure of surface layer of autoclave wall materials. Constr Mater Prod 1(2):4–10

Volokitin O, Volokitin G, Skripnikova N, Shekhovtsov V (2016) Plasma technology for creation of protective and decorative coatings for building materials. In: AIP conference proceedings, vol 1698, p 070022

10

Peculiarities of Phase Formation in Artificial Ceramic Binders for White-Ware Compositions

I. Moreva[(✉)], E. Evtushenko, O. Sysa, and V. Bedina

Belgorod State Technological University named after V G Shukhov, Belgorod, Russia
moreva_bstu@mail.ru

Abstract. The production of sanitary white wares traditionally uses multi-component mixes, which is necessitated by a whole complex of properties: high density and low humidity of molding slurries with low thixotropic strengthening and good filterability of the slurry. However, modern understanding of the structure and properties of materials open opportunities for optimization of technological process and production of higher-quality articles. The implementation of activation technologies and replacement of traditional molding slurries by artificial ceramic binders will reduce the number of components in mixes and optimize the production of white wares. Since the achievement of working performance of ceramic materials substantially depends on the phase formation processes in a material during sintering. Current work analyses the phase transformations occurring at all stages of white ware production as per proposed technology.

Keywords: Artificial ceramic binders · Phase conversions · Porcelain · Pottery Ceramic slurry

1 Introduction

Artificial ceramic binders (ACBs) are molding suspensions produced by the technology of high-concentration ceramic binding suspensions (Pivinskii 2003) from nonplastic materials (quartz sand, quartz glass, fire clay, etc.). The binding properties of such suspensions are determined not by clay minerals (as in the case of traditional ceramic slurries), but by a colloid component that forms during milling and which content can reach up to 5%. Previous studies have shown the possible production of ACBs on the basis of clay materials after preliminary thermal activation. The technology allows producing high-density molding suspensions with record-low moisture content: Wsusp = 16%, $\rho = 1950$ kg/m^3 (for traditional slurry 35% and 1740 kg/m^3, respectively). This technology is interesting for the production of white wares, because it allows unifying the properties of implemented raw materials and reduce the number of components in the paste from more than 7 down to 3. This work studies the peculiarities of ACB formation on the basis of thermally activated mixes of typical white ware composition: 50% of clay (kaolin from Prosyanoye deposit), 25% of fluxes (Vishnevogorsk spar) and 25% of filling aggregates (Ziborovsk sand).

2 Methods and Approaches

Initial materials were broken, mixed and wetted to 15–20% with consequent formation of bricks. After drying, the bricks were burnt at the temperature of 950–1000 °C with shock cooling for increased system activity. The thermally activated mix was milled as per the technology of highly concentrated ceramic binders for suspensions with introduction of electrolytes and liquid phase at the loading of the first part of the material. The XRD analysis of the investigated materials was made on DRON-3 diffractometer. The diffractograms were recorded with filtered CuKα-radiation (Ni filter); the tube voltage was 20 kV; tube anode current was 20 mA; the measurement limit was 10,000–4,000 pulses per second; the detector rotation speed was 2.4 min^{-1}; the angular mark was 10. The phases were identified using JCPDF cards. The samples with the dimensions of 30 × 30 × 30 mm from obtained suspensions and slurries were formed by molding into gypsum molds. The dried samples were burnt in a periodic kiln with silicon carbide heating elements at the temperature of 800–1200 °C with holding at maximum temperature for 5–10 min.

3 Results and Discussion

The preliminary thermal treatment of the studied raw materials allowed weakening crystal lattices of the minerals, creating the structural non-stability due to polymorph transformations of quartz (d/n, Å – 4.27; 3.35; 2.46; 2.29) and dehydration of clay and hydromicaceous minerals (d/n, Å – 7.23; 3.36; 4.48; d/n, Å – 10.16; 3.30; 2.91) (Fig. 1). After such a thermal treatment, a part of material passes into active state (Evtushenko et al. 2007).

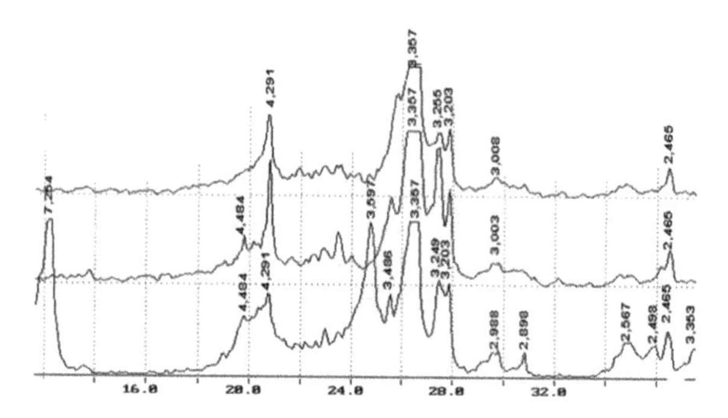

Fig. 1. Changed phase composition of studied mixes after activation and consequent milling: (a) initial untreated mix; (b) mix after thermal activation at 950 °C; (c) after wet milling.

The preliminary thermal activation of the material, consequent wet milling and high density of the samples facilitates the intensification of burning, in particular, appearance of mullite seeds (d/n, Å – 3.39; 5.42; 2.71; 2.55) beginning from 950 °C. In traditional compounds, the formation of mullite begins only at 1100 °C. A part of crystalline phases of the ACB in the white ware composition transits into the melt, which is witnessed by the appearance of amorphous phase and decrease in the intensity of reflections that are typical for quartz (d/n, Å – 3.35; 4.27; 2.46) and feldspars (d/n, Å – 3.24; 3.22) (Fig. 2).

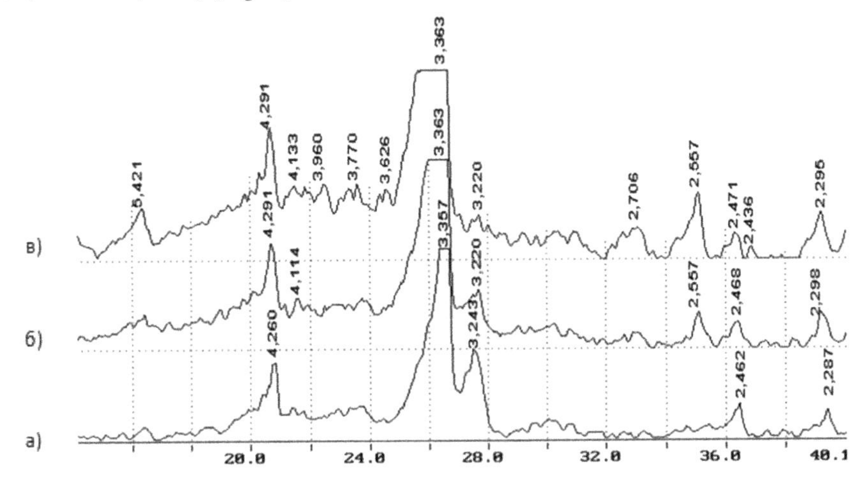

Fig. 2. X-ray diagrams of burnt samples of ACB in white ware composition: (1) 950 °C; (2) 1100 °C; (3) 1200 °C

Probably, a denser structure of the molding leads to an intense amorphization and sintering at temperatures close to 1100 °C, while at 1200 °C close-packed structures start to crystallize and stabilize. The high intensity of sintering and formation of a stronger structure of ACB samples is testified by gradual increase in the coefficient of crystallinity (Cc) with increased burning temperature: from 58.5% at 600 °C up to 91% at 1200 °C, while for cement plant slurry it is 66.5% at the final burning temperature. The high values of Cc for samples of ACB are probably connected with more intense processes of mullite and cristobalite formation.

4 Conclusions

Thus, the revealed peculiarities of phase formation in ACBs of white ware compositions demonstrate that the material has high reactivity in a wide temperature range. The preliminary thermal activation of raw materials with consequent sintering of the material facilitates earlier beginning of mullite and melt formation. The mentioned processes are determining for the formation of the strong structure of the material and high operation characteristics of the ceramic products.

Acknowledgements. The work is realized in the framework of the Program of flagship university development on the base of Belgorod State Technological University named after V.G. Shoukhov, using equipment of High Technology Center at BSTU named after V.G. Shoukhov.

References

Evtushenko EI, Sysa OK, Moreva IYu (2007) Controlling the properties of raw materials, casting slips, and pastes in fine-ceramic technology. Stroit Mater 8:16–17

Pivinskii YuE (2003) Highly concentrated ceramic binding suspensions (HCBS) and ceramic castables. Stages in research and development. Refract Ind Ceram 44(3):152–160

11

Structural-Phase Stabilization of Clay Materials in Hydrothermal Conditions

O. Sysa(✉), E. Evtushenko, I. Moreva, and V. Loktionov

Institute of Chemical Technology, BSTU named after V.G. Shukhov, Belgorod, Russia
sysa1975@inbox.ru

Abstract. The results of thermal and X-ray phase analysis of hydrothermal stabilized kaolin clay are given. An evaluation method has been suggested of crystalline structure order according to the strength degree of elementary contacts in clay suspensions. It has been noted that hydrothermal stabilization may result in crystallohydrate crystalline structure perfection, in saturation, extraction or rearrangement of water molecule in kaolinite clay structure, and influence new phase generation during ceramic material baking.

Keywords: Kaolin clay · Hydrothermal processing · Degree of sophistication . Crystalline · Structural-phase changes · Ceramics

1 Introduction

Structure imperfection often defines material quality and properties. The structure perfection for layered silicates is determined by the structure of aluminosilicate layers themselves and their positioning within crystallite. It is known that kaolinite perfect crystals are hexagonal plane particles of the regular shape which allows free sliding relative each other and providing plasticity, liquescence and close arrangement goods formation. Crystallite distortion results in distortion of kaolinite plates due to lineal and point defects. It deteriorates rheological behaviour of kaolinite suspensions, liquescency instability (Kukovskiy et al. 1966).

It is possible to affect clay material structure by several methods (natural, mechanical, chemical, biological and thermal) (Evtushenko et al. 2009). Hydrothermal clay treatment can be the most effective when it is accompanied by Rehbinder effect (adsorptive plastification) (Rebinder et al. 1972), which accelerates restructurisation and minimizes crystal defects in clay minerals for a short period of time (Evtushenko et al. 2006).

2 Methods and Approaches

X-ray phase material analysis has been done with a diffractometer "DRON-3". XRD-pattern was shot with CuKα filtered radiation, – radiation (Ni – filter); voltage across the tube is 20 kw; anode tube current is 20 mA; measurement range is 10000–4000 counts per second; detectors rotation rate is 2,4 rot/min; angular mark is 10. Card index JCPDF was used for phase identification of elementary contacts durability formed without particles in clay slurry. The calculation was done according to Ur'ev model (1972, 2002) for low aggregate suspensions (Zubehin et al. 1995):

$$F_1 = \frac{\gamma_m \eta_{\min} d^2}{6,4}, \quad (1)$$

where γ_m is share rate corresponding to complete structural destruction, s-1; η_{min} is effective viscosity corresponding to complete structural destruction, Pa s; d is average particle diameter, m.

Differential-thermal analysis was done with a derivatograph OD-102.

3 Results and Discussion

Kaolin clay of local deposits has been studied, which show structural changes of clay mineral after steaming in a pressure vessel at pressure from 1 up to 4 MPa.

With X-ray phase analysis we determined constant phase clay composition before and after treatment, with significant deviations of kaolinite crystalline structure. Hydrothermal modification at saturated steam pressure up to 4 MPa drives up crystallinity index according to Hinckley and intensivity of main diffractional kaolinite reflections (Table 1), that testifies to a greater mineral structure order in the treated raw material (Shlykov et al. 2006).

But "Hinckley crystallinity index" C_h decrease depending on defects in layer composition. This parameter is not always applicable to index correlation of index characteristics with its real structure.

Basing on the said above the authors developed an estimation method of material structure sophistication degree according to the strength magnitude of elementary contact, formed between particles in clay suspensions.

Table 1. Change in "Hinckley crystallinity index" (Ch) of kaolinite after hydrothermal treatment

Kaoline clay deposits	Initial kaoline	Kaoline after pressure treatment 1,6 MPa	Kaoline after pressure treatment 4,0 MPa
Zhuravliniy Log	0,20	0,22	0,46
Gluhovetskoe	0,77	0,83	1,27
Kyshtymskoye	0,55	0,47	0,61
Prosyanovskoye	0,51	0,63	0,63
Novoselitskoye	0,26	0,35	0,31

It has been found that elementary contact strength decreases at hydrothermal treatment pressure and temperature increase by more than an order. Ranging by crystalline structure defects these kaolines can be placed as follows: Zhuravliniy Log > Glukhovetskiy > Kyshtymskiy > Prosyanovskiy and Novoselitskiy.

Some differences in endo-and exothermal processes during baking in kaoline samples depending on hydrothermal modification have been studied (Gorshkov et al. 1988). It has been determined that maximum endothermic effect of kaoline dehydration shifts towards higher temperatures that testifies to crystalline structure sophistication. Restructurisation results in additional hydration and lessening connection energy of crystal water in kaoline, facilitates water molecule extraction penetrating into basis of tetrahedral kaoline layers.

Thermal capacity change dependence of the studied materials during baking has been studied. In the temperature range 950–1020 °C there are several exothermic extreme values testifying to crystallization possibility of a variety of phase (β-crystobalite, mullite, sillimanite, γ-alumina). Structural change of the initial material causes temperature shifts at the initial crystallization phase, as in clays with clearly seen crystalline structure mullite is formed at lower temperature and in greater quantity than from disordered structure minerals.

4 Conclusions

Hence it has been determined that hydrothermal treatment improves significantly stabilization of clay minerals structure as well as ceramic material baking processes, that has been proved by several research methods. Due to crystallohydrate structure change, saturation, water molecule distribution in kaolinite structure dehydration parameters change, processes take place which cause temperature shift of new phase chilling point in the interval of lower temperatures.

Acknowledgement. The work is realized in the framework of the Program of Flagship University development on the base of the Belgorod State Technological University named after V.G. Shukhov, using equipment of High Technology Center at BSTU named after V.G. Shukhov.

References

Evtushenko EI, Sysa OK (2006) Structural modification of clay raw material in hydrothermal conditions. University news. North-Caucasian region. Technical sciences series, no 2, pp 82–86

Evtushenko EI, Sysa OK, Moreva IYu, Bedina VI, Trunov Y (2009) Raw material preparation refining during activation processes in ceramics technology. Glass Ceram 1:15–16

Gorshkov VS, Saveliev VG, Fedorov NF (1988) Physical chemistry of silicates and other high-melting compounds. High School, Moscow

Kukovskiy Y (1966) Structure peculiarities and physic-chemical properties of clay minerals. Naukova Dumka, Kiev

Rebinder PA, Ur'ev NB, Shukin Y (1972) Physic-chemical mechanics in chemical technology of disperse systems. Theor Bases Chem Technol 6:16–24

Shlykov VG (2006) X-ray analysis of disperse soils mineral composition. GEOS, Moscow

Zubehin AP, Strakhov VI, Chekhovskiy VG (1995) Physico-chemical research methods of high-melting non-metal and silicate materials. Synthesis, Saint-Petersburg

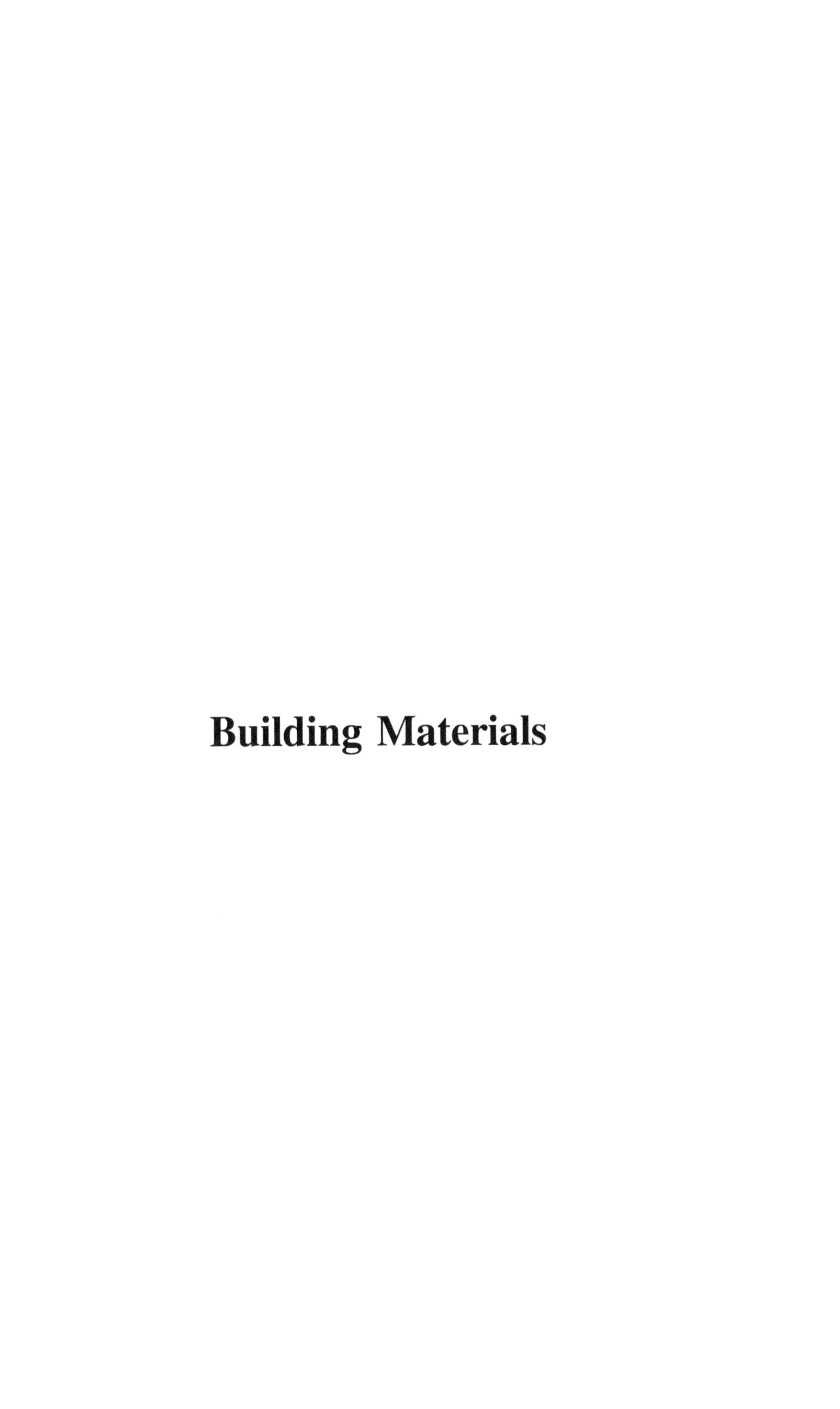

Building Materials

12

Impact of Thermal Modification on Properties of Basalt Fiber for Concrete Reinforcement

V. Strokova, V. Nelyubova(✉), I. Zhernovsky, O. Masanin, S. Usikov, and V. Babaev

Department of Materials Science and Materials Technology, Belgorod State Technological University named after V.G. Shukhov, Belgorod, Russia
vvnelubova@gmail.com

Abstract. The paper shows feasibility and efficacy of thermal modification of basalt fiber to increase its corrosion resistance and durability in a cement matrix. The authors justify the mechanism of phase and structure transformation of the fiber subsurface layer providing its increased physicochemical properties.

Keywords: Basalt fiber · Fiber · Thermal modification · Oxidation

1 Introduction

Under the conditions of real operation, the elements of concrete structures suffer cyclic, fatigue, impact, stretching and twisting loads, which causes uncontrolled cracking and consequent destruction of the cement matrix (Lesovik et al. 2018).

This problem was solved by using fibers to reinforce concrete matrix, which prevents brittle fracture of concrete and enables control of cracking (Klyuyev et al. 2013).

The analysis of approaches to enhancing the corrosion resistance of glass fiber in alkaline medium demonstrates that the thermal treatment of basalt fiber without its crystallization with initiation of a number of processes that increase alkali resistance and strength of the fibers should be considered as the most advanced and economically feasible method for increasing the stability of basalt fiber for its consequent application for concrete production on the basis of cement binders. (Knot'ko et al. 2007; Knotko et al. 2011)

According to the accepted working hypothesis, the increase of alkali resistance of fiber surface should be achieved by thermal treatment. This method is considered to be the simplest, affordable and economically reasonable. Its technical performance is predetermined by a range of physicochemical processes taking place in glass fibers during thermal treatment: oxidation, structure and diffusive rearrangement of ions in the material, annealing and densification of the structure, pre-crystallization and crystallization, etc.

2 Methods and Approaches

The main component in the work was basalt fiber produced at Novgorod glass fiber plant using machine factory BASK by blowing the melt by vertical air stream.

The method for investigation of thermal treatment impact on the fiber properties included step-wise heating of the fiber from 300 to 700 °C (700 °C is the working temperature of the basalt fiber, T_{work}) with the step of 100°. The isothermal holding time was 30 min, which considered the microscopic dimensions of fibers and striving to provide high productivity of the thermal treatment process. The fiber was cooled down in air by convective heat exchange. The fiber specimens then were tested for alkali resistance in model solutions represented by cement water extract.

3 Results and Discussion

According to preliminary data on the resistance of the fiber from different producers in alkaline and acidic medium, the chosen basalt fiber stands out for its high alkali and acid resistance. This is explained by the wash-out of cations persisting in vitrified phase sue to alkaline hydrolysis into the silicon dioxide solution with formation of aluminates and zincates with consequent liberation of anions SiO_4^{4-}, $Si_2O_5^{2-}$ and SiO_3^{2-}. Insoluble complex aluminate and zincate salts accumulate on the chemically active fiber surface, which inhibits further leaching of silicon. In the absence of further introduction of alkaline component for supporting necessary pH, the decomposition retards.

The corrosion of basalt fiber and its consequent disruption is caused by both interaction with hydroxyl groups of silicon-oxygen radicals and capability of cations dissolved in water alkaline medium to exchange with cations comprising the basalt fiber.

Noteworthy, during the utilization of basalt fiber in real conditions in concrete, the fiber dissolution degree will be not that significant, since the solution processes will die down along with curing and solidification of the cement. Nevertheless, the chemical processes involving the fiber that take place in concrete during its life should not be underestimated. This necessitates the development of the fiber modification method to improve its resistance to aggressive alkali medium.

The analysis of obtained data (Fig. 1) confirmed that thermal treatment of fiber at 500 °C was the most effective, as it promoted its alkali resistance as compared to non-treated fiber by 35.3% (after 28 days). Further increase of the treatment temperature was useless because, firstly, the fiber alkali resistance reduced; secondly, it was not economically reasonable. The effect of increased alkali resistance at 300 and 400 °C amounted to 26.7 and 30%, respectively.

Heating of basalt fibers leads to negligible reduction of their mass due to loss of adsorbed and chemically bound water, while further heating increases the mass by binding of air oxygen during divalent iron FeO oxidation into trivalent Fe_2O_3. The increased oxidation degree of iron cations leads to their decreased ion radius and reduction of the coordination number with formation of groups of iron-oxygen tetrahedrons $[FeO_4]^-Na^+$ that are embedded into the glass structure forming complex iron-aluminum-silicon-oxygen framework and increasing the degree of association, stability

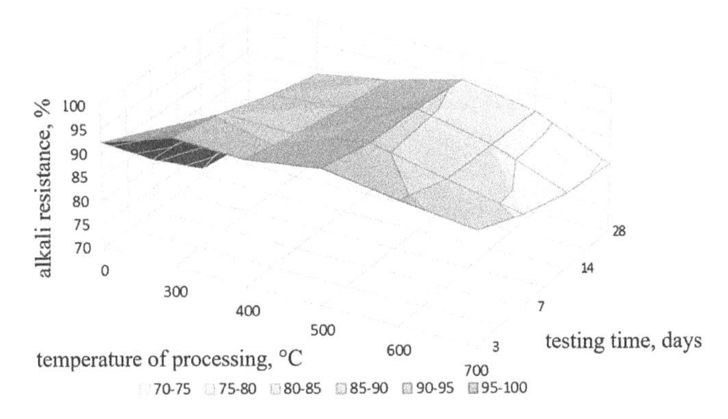

Fig. 1. Alkali resistance of fiber versus preliminary treatment temperature and testing time

and density of the structure. Similar structural restructuring triggers the diffusion redistribution of modifying cations (Na^+, Ca^{2+}) in the fiber.

The thermal treatment is accompanied by microdefect healing, structure densification and its approaching to stable equilibrium state.

The processes described above are most active at 500 °C—the temperature comparable with the glass transition temperature T_g (corresponds to the viscosity of $10^{12,3}$ Pa s and transition of glass from solid into plastic state)—which conditions the highest alkali resistance of fibers and stability of their work in cement concretes.

At higher temperatures (600–700 °C) and decreased viscosity (10^{10}–10^7 Pa s), the surface of basalt fibers suffers the manifestation of structural pre-crystallization and crystallization processes accompanied by the formation of diverse defects, which increase the free energy of the material and makes it more chemically active. This results in accelerated interaction with alkali and increased fiber mass losses. Besides, the thermal treatment at high temperature leads to the strength loss of basalt fibers: at 600 °C to negligible strength decrease; at 700 °C to the loss of 50% of initial strength.

Thermal treatment of basalt fibers at 500 °C initiated a number of important physicochemical and structural processes leading to the modification of glass fiber surface with acquisition of higher chemical resistance.

The holding close to the glass transition temperature T_g leads to gradual densification of the structure and its transition into more stable equilibrium state with higher chemical resistance. During the structure stabilization, numerous microdefects heal inside and on the surface of the fiber, which results in its decreased free energy and chemical activity. The structure densification also decreases the rate of numerous diffusion processes, which are the basis of any chemical reactions with solid phases.

Activation of iron ion oxidation during thermal treatment leads to substantial restructuring of the surface layer of basalt fiber: modifying ions of Fe^{2+} after oxidation to Fe^{3+} embed as iron-oxygen tetrahedrons $[FeO_4]^{5-}$ into aluminum-silicon-oxygen lattice of the glass, thus increasing the degree of its association and chemical resistance.

Preliminary 30-min processing of the basalt fiber at 500 °C by 35% increases its resistance in alkaline medium of curing cement and can be recommended as effective, simple and economically reasonable method for modifying basalt fiber for its effective implementation as a micro-reinforcing component of fibrous concretes.

4 Conclusions

Thus, the processes initiated by thermal treatment promote the corrosion resistance of fibers in alkaline medium, which prolongs the corrosion resistance and increases the efficacy of fiber application as a micro-reinforcing component for cement composites.

Acknowledgements. The work has been fulfilled within the project Federal Target Program of Research and Development on "Priority Development Fields of science and technology sector in Russia for 2014–2020", the unique project number is RFMEFI58317X0063.

References

Klyuyev SV, Klyuyev AV, Sopin DM, Netrebenko AV, Kazlitin SA (2013) Heavy loaded floors based on fine-grained fiber concrete. Mag Civil Eng 38(3):7–14

Knot'ko AV, Garshev AV, Davydova IV, Putlyaev VI, Ivanov VK, Tret'yakov YuD (2007) Chemical processes during the heat treatment of basalt fibers. Prot Metals 43(7):694–700

Knotko AV, Pustovgar EA, Garshev AV, Putlyaev VI, Tret'yakov YD (2011) A protective diffusion layer formed on surface of basaltic fiberglass during oxidizing. Prot Metals Phys Chem Surf 47(5):658–661

Lesovik VS, Glagolev ES, Popov DY, Lesovik GA, Ageeva MS (2018) Textile-reinforced concrete using composite binder based on new types of mineral raw materials. In: IOP conference series: materials science and engineering, vol 327, no 3

13

Designing High-Strength Concrete using Products of Dismantling of Buildings and Structures

T. Murtazaeva[1(✉)], A. Alaskhanov[1], M. Saidumov[1], and V. Hadisov[1,2]

[1] Millionshchikov Grozny State Oil Technical University, Grozny, Russia
tomamurtazaeva@mail.ru
[2] Ibragimov Complex Research Institute, RAS, Grozny, Russia

Abstract. The paper presents an analysis of experience of using the products of dismantling of buildings and structures, the technology of recycling of secondary raw materials to produce secondary raw materials for concrete. We presented results of tests of heavy concrete based on filled binders using the products of processing of concrete and brick scrap.

Keywords: Building demolition products · Concrete scrap · Brick scrap · Ecology · Recycling · Secondary aggregate · Fine ground aggregate filled with binder

1 Introduction

A great interest for application of the secondary product of crushing concrete scrap of dismantling buildings and structures, according to the authors of the paper, is related to the possibility of its application as a fine-milled mineral component in filled binders, based on the use of which it is possible to produce high-strength concrete, including monolithic high-rise construction (Bazhenov et al. 2011; Batayev et al. 2017; Murtazaev et al. 2009).

2 Methods and Materials

The following materials were used as raw materials for concrete: natural sand from the Chervlenskoye deposit, crushed stone from gravel of 5–20 mm fractions from the Argunsky and Sernovodsky deposits, imported crushed stone of a 5–20 mm fraction from granite-diabase rocks of the Alagirsky deposit of the Republic of North Ossetia-Alania, local non-additive portland cement of brand PC 500 D0, plastification additives Polyplast and D-5.

3 Results and Discussion

To obtain the optimal formulations of high-strength concrete (HSC) with the integrated use of local raw materials, including of technogenic nature, compositions of filled binders (FB) with fine-milled mineral filler of technogenic nature (MFTN) of HB-75:25 and HB-60:40grades, allowing to obtain high-strength cement stone with noticeably smaller pores and less shrinkage (Murtazaev et al. 2009; Udodov 2015; Murtazaev and Salamanova 2018).

From the test results of HSC based on FB, it can be seen that the dynamics of the strength of concrete on FB are noticeably different from the dynamics of the growth of concrete strength on Portland cement.

It was established that the process of durability of concrete on FB at an early age (1–3 days) is accelerated by 1.5–2 times. So, concrete on FB at the age of 1 day has a strength of about 33–36% of the designed, and 3 days old - this indicator reaches 70%. 7-day strength of concrete, produced using FB, is about 85–90% of the designed, which is significantly higher than traditional compositions on ordinary Portland cement. These indicators for concrete on Portland cement at the age of 1, 3 and 7 days are about 24, 35 and 70% of the designed strength, respectively.

4 Conclusions

The optimal formulations of highly mobile concrete mixtures were designed using local natural and technogenic raw materials with a grade of P5 cone sediment and persistence for more than 8 h to obtain HSC classes of compressive strength up to B60-B80 with unique performance properties.

References

Bazhenov YM, Bataev DK-S, Mazhiev KN (2011) Fine-grained concretes from recycled materials for the repair and restoration of damaged buildings and structures. FE "Sultanbegova Kh.S.", Grozny, 342 p

Batayev DK-C, Saidumov MS, Murtazaeva TS-A (2017) Recipes of high-strength concretes on technogenic and natural raw materials. In: Materials of the All-Russian Scientific and Practical Conference Dedicated to the 60th Anniversary of the Building Department of Millionshchikov GSTU, 12–13 October 2017, Bisultanova P.Sh., Grozny, pp 109–117

Murtazaev S-AY, Bataev DK-S, Ismailova ZK (2009) Fine-grained concretes based on fillers from secondary raw materials. Comtechprint, Moscow, 142 p

Udodov SA (2015) Re-introduction of plasticizer as a tool for controlling the mobility of concrete mix. In: Proceedings of the Kuban State Technological University, no 9, pp 175–185

Murtazaev SAY, Salamanova MS (2018) Clinker-free binders based on finely dispersed mineral components. In: Collection: Conference Proceedings, pp 707–714

14

Regularities in the Formation of the Structure and Properties of Coatings Based on Silicate Paint Sol

V. Loganina(✉), E. Mazhitov, and V. Demyanova

Department "Quality Management and Technology of Construction Production", Penza State University of Architecture and Construction, Penza, Russia
loganin@mail.ru

Abstract. The authors established a higher quality of the appearance of coatings based on silicate paint sol in comparison with coatings based on silicate paints. Information is provided on the regularities in the formation of the quality of the appearance of coatings on the basis of silicate paint sol is that the silicate paint sol has a higher value coefficient of the wetting and spreading on the cement substrate in comparison with silicate paint. It was established that when using a paint brush on the cement substrate, there has been some slowing of the rise time of the structure of silicate paint compositions based on polysilicate solutions. A higher value of the adhesion work based on polysilicate solutions were indentified, indicating the strong force of the paint and the cement substrate.

Keywords: Sol silicate paint · Polysilicate binder · Coatings · Wetting · Coating quality

1 Introduction

The problem of reliability and durability of protective decorative coatings of exterior walls of buildings is one of the topical scientific and technical objectives in materials science. It is known that the longevity of coatings depend on the type of binder, the technology of applying the paint composition, operating conditions, etc. (Ailer 1982). In the practice of finishing works, silicate paints, which are a suspension of pigments and fillers in liquid glass of potassium, proved to be very useful. To improve the performance of coatings based on silicate paints, it is important to develop methods for modifying liquid glass. Analysis of patent and scientific-technical literature shows that one way of modification is the introduction of organosilicon and other polymeric compounds into the binder. In works (Figovsky et al. 2012) there is an increased durability of silicate coatings when imposing of polymeric compounds. It is of interest to use polysilicates in silicate paints as film-forming substances that provide higher performance properties of coatings. However, at present time the questions of the formation of the structure and properties of coatings based on sol silicate paints have not been studied, the questions of mechanism for improving the operational properties of coatings based on them have not been considered. Polysilicates are characterized by a broad range of the degree of polymerization of anions and are dispersions of colloidal silica in an aqueous solution of alkali metal silicates. We have established a paint composition based on a polysilicate binder obtained by mixing liquid glass with a silica sol (Loganina et al. 2018a, b). It was found, that coatings based on polysilicate solutions are characterized by faster curing. Films based on polysilicate solutions have higher cohesive strength. The tensile strength of a film based on potassium liquid glass is Rp = 0.392 MPa, and the tensile strength of a film based on a polysilicate solution (15% Nanosil 20) is 1.1345 MPa. The paint forms a coating characterized by a uniform homogeneous surface.

To study the regularities in the formation of the quality of the appearance of coatings on the based on sol silicate paint, the character of filling on a porous cement substrate was considered.

2 Methods and Approaches

The character of the filling of the sol of the silicate paint was evaluated. The method of determining filling consisted in applying five parallel strips of paint and determining the degree of spreadability according to the number of stuck bands. Paint with an operating viscosity was applied to a glass plate measuring 200 × 100 × 1.2 mm. The spreading of five parallel strips was evaluated on a ten-point scale of filling.

The surface tension of the paint was determined by the drop method (stalagmometric method). Work of adhesion of the paint to a cement substrate was calculated using the Dupre - Young thermodynamic equation. The wetting operation and spreading coefficient were determined.

The quality of the appearance of the coatings was estimated from the surface roughness Ra by the method of scanning probe microscopy (SPM) (Chizhik and Syroezhkin 2010).

3 Results and Discussion

Previously, the rheological type of the solutions was determined. The rheological properties were evaluated by the indicators of conventional viscosity according to B3-4, critical shear stress with instrument Reotest-2. It is found, that all systems are typical pseudoplastic bodies. In the region of slow flow, the viscosity of polysilicate solutions gradually declined with increasing shear stress.

Analysis of data (Table 1) shows, that silicate paints based on polysilicate solutions have a long filling time. Thus, the time for bottling for a paint based on liquid glass is 6 min, and for a paint based on a polysilicate solution - 8 min 40 s. The degree of filling is satisfactory (no more than 10 min). For paints based on a polysilicate solution, a large work of adhesion to the substrate is characteristic. So, work of the adhesion of paint based on the potassium polysilicate solution to the substrate is 108.17 mN/m, while work of the adhesion of the paint based on potassium liquid glass is 96.82 mN/m. A higher value of work of the adhesion of paint based on polysilicate solutions indicates a stronger interaction of the paint and the cement substrate.

Table 1. Test results

The name of indicators	*Name of the paint composition*	
	Based on potassium liquid glass	*Based on potassium polysilicate solution*
Surface roughness, Ra, [μm]	16,208	10,880
The contact angle of wetting	50,9	51,6
Surface tension of the paint composition, [mN/m]	59,38	66,73
Filling the colorful composition *	7 min 40 s / 9	8 min 40 s / 9
Adhesion work, [mN/m]	96,82	108,17
Wetting operation, [mN/m]	37,44	41,44
Cohesion work, [mN/m]	118,76	133.46
Coefficient of spreading [mN/m]	−21,94	−25,29
Coefficient of wetting	0,815	0,81

Note: * Above the line are the values of time of restoration of the paint structure, below the line - the value of filling

The work of wetting paints based on polysilicate solution is higher, which indicates better wetting of the paint surface of the cement substrate. Thus, the work of wetting the sol of silicate paint on the basis of potassium polysilicate solution is 41, 44 mN/m, and on the basis of potassium liquid glass - 37.44 mN/m. When the sol of the silicate paint is applied to the substrate, the wetting and spreading coefficient increases, which indicates more favorable conditions for the formation of the quality of the appearance. The surface roughness of the coating based on silicate paint is Ra = 16.208 μm, and based on the potassium polysilicate solution, Ra = 10.880 μm. The quality of the appearance of the surface of the coatings formed by the sol with silicate paint, in accordance with GOST 9.032-74 ** "Unified system for protection against corrosion and aging. Coatings for paint and varnish, Groups, technical requirements and designations" is graded IV class, and on the basis of liquid glass - V class.

Testing of solution samples, colored with sol by silicate paint, was carried out for frost resistance by alternating thawing and freezing. Appearance of coatings was assessed according to GOST 6992-68. "Coatings for paint and varnish. Test method for resistance to atmospheric conditions". It was found, that coatings on the basis of the developed composition had withstood 40 test cycles, while the coating condition after 40 test cycles was estimated at I.1 points, which corresponds to the coating condition with no color change, chalking, mud retention.

To assess the waterproof properties of coatings, tests were carried out of solution samples stained with silicate and sol silicate paints. After curing of the coatings, water absorption was determined upon capillary suction. It was found, that water absorption by capillary suction of samples stained with silicate paint is 4.4%, and stained with silicate paint - 4.6%. The lower value of water saturation of samples colored with sol by silicate paint indicate a change in the pore size in the coating structure as compared to the coating based on silicate paint.

Higher waterproof properties of coatings based on sol silicate paint are caused, in our opinion, by the structure of the coating. Scanning probe microscopy (SPM) methods were used to estimate the local structure of the coating surface. It is established, that the surface of coatings based on potassium liquid glass contains a certain number of pores of the nanometric range, differing in size and shape. The maximum pore size is 19.8 μm. Pores with a diameter of 18.85 to 19.6 μm are mainly present, whereas in the coating based on the potassium polysilicate solution there are two groups of pores: from 19.25 to 19.8 μm and from 20.0 to 20.6 μm. The value of the maximum pore size is 21.2 μm. In the coating based on the polysilicate solution is observed a more uniform pore size distribution.

Coatings based on the developed paint are characterized by high adhesion (1.1–1.3 MPa), coefficient of vapor permeability - 0,00878 mg/m * hPa.

4 Conclusions

The properties of the paint and coating based on it meet the requirements for coatings for exterior decoration of buildings, have higher adhesion, sufficient vapor permeability.

References

Ailer P (1982) The chemistry of silica (Transl. from English), Part 1. Mir, Moscow, p. 416

Chizhik SA, Syroezhkin SV (2010) Methods of scanning probe microscopy in micro- and nanomechanics. Instr Meas Methods 1:85–94

Figovsky O, Borisov Yu, Beilin D (2012) Nanostructured binder for acid-resisting builder materials. J Sci Israel-Technol Advant 14(1):7–12

Loganina VI, Kislitsyna SN, Mazhitov YB (2018a) Development of sol-silicate composition for decoration of building walls. Case Stud Constr Mater 9:e00173

Loganina VI, Kislitsyna SN, Mazhitov YB (2018b) Properties of polysiylate binders for sol-silicate pains. Adv Mater Res 1147:1–4

15

Geopolymerization and Structure Formation in Alkali Activated Aluminosilicates with Different Crystallinity Degree

N. Kozhukhova[1(✉)], V. Strokova[1], I. Zhernovsky[1], and K. Sobolev[2]

[1] Belgorod State Technological University named after V.G. Shukhov, Belgorod, Russia
kozhuhovanata@yandex.ru
[2] University of Wisconsin-Milwaukee, Milwaukee, USA

Abstract. The work presents the results of grain-size analysis of alkali-activated aluminosilicate suspensions with different crystallinity degree. It is found that the crystallinity degree of aluminosilicate is inversely proportional to its solvability in strong alkaline substance. The mechanism of geopolymeric system formation during the geopolymerization process has been suggested.

Keywords: Aluminosilicates · Crystallinity degree · Structure formation · Geopolymerization

1 Introduction

The application of colloid and nano-sized silicate and aluminosilicate components for the synthesis of effective binding systems is one of the most attractive directions in the science of construction materials (Vivian et al. 2017; Dmitrieva et al. 2018; Sobolev 2016).

The earlier studies (Shekhovtsova et al. 2018; Galindo Izquierdo et al. 2009) discussed various factors influencing the ability of alkali activated cements to form the aluminosilicate structures from the anthropogenic aluminosilicates, in particular fly-ash (Kozhukhova et al. 2018; Wang et al. 2018).

2 Materials and Methods

To estimate the viability of the research hypothesis three types of natural aluminosilicates with a different crystallinity degree were used:

- Obsidian - effusive rock of an acidic composition and amorphous structure;
- Pearlite - effusive rock of an acidic composition and crypto-crystalline structure;
- Crouan – intrusive compound acidic rock with a hollow crystalline structure.

To prepare alkaline silicate suspensions the equal volumes (50 g) corresponding to each sample of milled aluminosilicate material were placed into glass bottles and mixed with 50% NaOH water solution.

These suspensions were mixed for three days (72 h) using a LS-110 mixing device. The specific surface and the average size of grains of aluminosilicate powders was performed using a laser analyzer ANALYSETTE 22 NanoTec plus.

3 Results and Discussions

This report is based on the hypothesis that during the alkaline activation of aluminosilicate particles the dissolving process is gradual, starting with the dissolution of surface layers. As the result, the alkali aluminosilicate gel is formed which acts as a binding base for further geopolymerization. At the same time, the aluminosilicate component crystallinity degree influences its solubility in the alkaline medium.

To test the research hypothesis the average size of the aluminosilicate particles average size was determined as well as the specific area in the initial condition and also after the alkaline activation (Table 1).

Table 1. The particle size and specific area of alumosilicate powders after the alkaline activation

№	Aluminosilicate type	Average particle size, µm		Change, %	Specific area, m^2/kg		Change, %
		Before activation	After activation		Before activation	After activation	
1	Crouan	11	9	–16	910	1003	10
2	Perlite	11	14	21	838	774	–7
3	Obsidian	14	17	23	785	642	–18

The results of grain size analysis for crouan after the alkaline activation prove that the average particle size is reduced relative to the particle size before activation. It may be caused by crouan grains subsolution resulting in the reduction of the particle size.

At the same time in the alkaline activated suspensions of perlite and obsidian there is a tendency of particle size increase in comparison with those before activation.

Hence, for crouan there is an increase of the specific area of the solid phase after the activation and a decrease of average particle size.

In case of perlite and mostly obsidian, the alkaline activation has a reverse effect: the specific area decreases and the average particle size increases (Fig. 1).

The received data of the grain size analysis suggested a scheme of aluminosilicate structure formation that occurs during the geopolimerization. This scheme includes two simultaneous processes: the dissolution of the aluminosilicate component and the formation of the alkali aluminosilicate gel, which is the chemical interaction of alkali aluminosilicate gel with unreacted grains. The chemical interaction causes the formation of a «gel layer – unreacted grain» in investigated system.

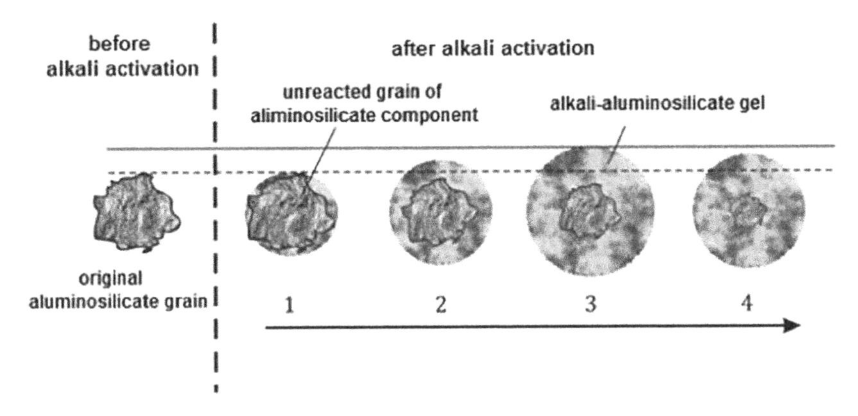

Fig. 1. The structure formation mechanism in the system "gel layer – unreacted grain of aluminosilicate component" during the alkali activation.

The lower aluminosilicate crystallinity degree results in a high intensity of the dissolution of aluminosilicate particles and a thicker surface gel layer based on the newly formed compound.

4 Conclusion

The crystallinity degree of aluminosilicates is inversely proportional to the reactivity in alkali systems, which is controlled by the solubility degree in highly alkali medium. Based on this observation, it was suggested that the geopolimerization scheme in the system «gel layer – unreacted grain of aluminosilicate component» occurs during the alkaline activation.

Acknowledgements. The work has been fulfilled within the project Federal Target Program of Research and Development on "Priority Development Fields of science and technology sector in Russia for 2014–2020", unique project number is RFMEFI58317X0063.

References

Dmitrieva TV, Strokova VV, Bezrodnykh AA (2018) Influence of the genetic features of soils on the properties of soil-concretes on their basis. Constr Mater Products 1:69–77

Galindo Izquierdo M, Querol X, Davidovits J, Antenucci D, Nugteren HW, Fernández-Pereira C (2009) Coal fly ash-slag-based geopolymers: microstructure and metal leaching. J Hazard Mater 166(1):561–566

Kozhukhova NI, Zhernovsky IV, Sobolev KG (2018) Effect of variations in vitreous phase of low-calcium aluminosilicates on strength properties of geopolymer systems. Bull BSTU named after V.G. Shukhov 3:5–12

Shekhovtsova J, Zhernovsky I, Kovtun M, Kozhukhova N, Zhernovskaya I, Kearsley PE (2018) Estimation of fly ash reactivity for use in alkali-activated cements - a step towards sustainable building material and waste utilization. J Cleaner Prod 178:22–33

Sobolev K (2016) Modern developments related to nanotechnology and nanoengineering of concrete. Front Struct Civil Eng 10(2):131–141

Vivian F-I, Pradoto R GK, Moini M, Kozhukhova M, Potapov V, Sobolev K (2017) The effect of SiO_2 nanoparticles derived from hydrothermal solutions on the performance of portland cement based materials. Front Struct Civil Eng 11(4):436–445

Wang YS, Provis JL, Dai JG (2018) Role of soluble aluminum species in the activating solution for synthesis of silico-aluminophosphate geopolymers. Cement Concrete Compos 93:186–195

16

Structurization of Composites when using 3D-Additive Technologies in Construction

M. Elistratkin[1(✉)], V. Lesovik[1], N. Chernysheva[1], E. Glagolev[2], and P. Hardaev[3]

[1] Department of Building Materials, Products and Designs, Belgorod State Technological University named after V.G. Shukhov, Belgorod, Russia
elistratkin.my@bstu.ru, mr.elistratkin@yandex.ru

[2] Department of Construction and Municipal Economy, Belgorod State Technological University named after V.G. Shukhov, Belgorod, Russia

[3] Department Industrial and Civil Engineering, East Siberian State University of Technology and Management, Ulan-Ude, Russia

Abstract. One of new and perspective lines of development in the field of construction technologies is the integration of additive production elements. The laboratory study of these issues revealed critical challenges that slightly slowed down the introduction of construction printing in daily construction practice. One of such problems is a big difference of structurization conditions of additive composites produced via traditional methods. The paper provides the analysis of factors exerting negative influence on structurization and considers the possibilities of solving such challenges.

Keywords: Construction printing · Mixing ratio · Structurization of composites · Composite binding agents · Mineral additives

1 Introduction

One of new and quite perspective development areas of construction technologies is the integration of additive production elements. The active development of construction printing launched a decade ago allowed creating some concepts demonstrating only a mere part of its potential and drawing public attention and investments to its development (De Schutter et al. 2018). However, it also caused some critical problems that slightly slowed down the introduction of construction printing in daily construction practice.

First of all, such problems include structural reinforcement. In the most cases traditional reinforcement methods (frames, rods, grids) do not correspond to the ideology of additive production, which implies that the construction robot installs the building structure without too much human involvement and use of various additional technical tools. Various methods of dispersed (Christ 2015) and textile reinforcement (Lesovik et al. 2017) may possibly solve the matter, and in the long-term perspective this problem may be solved by making the construction printer place or 'raise' the reinforcement according to the design project.

Another problem, which is currently being tackled by scientists, is the production of efficient mixes for printing, which is considered controversial from the perspective of the traditional concrete technology and with regard to its sufficient properties. It is impossible to ensure competitiveness of construction additive technologies in relation to traditionally applied methods without the development of new principles of their creation that would integrate classic approaches and the latest achievements in construction materials science (Yi et al. 2017).

2 Methods and Approaches

The study used the traditional cement-sand mortar at the ratio of 1:4 at W/C = 0.45… 0.5 as an extrusion printing mix. The following refer to special properties of mortar, which make its different from standard mixes:

- ability to easily pass through the extruder and a 20 × 20 mm nozzle without losing uniformity and sticking to walls;
- ability to hold its shape after extrusion and resist loading of at least 5 layers placed on top without intermediate curing.

The specified qualities were obtained due to combination of two Russian additives characterized by availability, low cost and good technological effectiveness. The used additives allow receiving various mixes for construction printing at the C:S ratio from 1:3 to 1:5 that preserve special properties within 25…30 min.

3 Results and Discussion

The printing with the developed mixes on a laboratory unit (Fig. 1) sets the task to obtain the flattest surface without post-processing.

Fig. 1. Printing with developed mixes on a laboratory unit

Table 1 shows the strength properties of the mix during processing in various conditions. The preparation of a molding compound is critical since this stage ensures the formation of traditional and special properties.

According to our experience in the study of construction printing, it is more preferable for industrial facilities to ensure continuous supply of the mix into a small bin feeder installed on a forming device.

Table 1. Strength of mixes (MPa) depending on curing conditions

Composition	1 day	3 day	7 day	28 day
1:3 (in water)	3.7	8.2	12.6	17.9
1:3 (in insulation)	4.2	8.2	11.3	14.3
1:3 (in air)	4	6.9	8.1	8.5
1:4 (in water)	2.9	6.3	9.4	12.9
1:4 (in insulation)	3	6.8	10.5	15.2
1:4 (in air)	3	5.4	6.6	6.8

The stage when the mix passes through the extruder is followed by its additional mixing, decrease in viscosity, and utilization of some amount of air. Printing with slight premolding of lower layers with the newly placed ones ensures the formation of stronger contact between them compared to free mix outflow, but at the same time imposes increased requirements on the ability of a mix to keep its shape after extrusion (Secrieru et al. 2017).

The third stage is characterized by the largest duration and ensures the formation of final indicators of the additive structure.

The curing conditions of composites in the printed structure significantly (for the worse) differ from traditional methods of concrete works (Zharikov et al. 2018). The wall assemblies made via outline printing have small effective sectional area at their quite big surface area, which leads to their fast dehydration. Table 1 shows the influence of curing conditions on strength accumulation velocity.

It is less likely possible to create favorable conditions for structurization of composites due to external moistening, therefore it is possible to define solution to this problem: reduction of a binder's demand in water and the maximum increase in early strength during the period until the main amount of liquid has not evaporated yet.

As it was noted in some works (Chernysheva et al. 2013; Sumskoy et al. 2018; Lesovik et al. 2014; Kuprina et al. 2014; Elistratkin et al. 2018), a good solution in such cases may be the replacement of portland cement with composite binding agents that contain mineral additives with developed microporosity (zeolites, utilization products of some ceramic construction materials). Such products are able to create some moisture stock, which maintains hydration for some time needed for the material to become strong.

4 Conclusions

Alongside with various processing technologies, quite often complicating the design of a construction printer, the essential positive effect may be achieved by using composite binding agents with required properties: required rheology, ability to quickly gain strength during solidification in the conditions of fast dehydration, reduced or zero shrinkage. Such measures will provide for a simpler solution to the task of creating favorable conditions for composite structurization in construction printing thus establishing a self-sufficient system not too much dependent on external factors.

Acknowledgements. The study is implemented in the framework of the RFBR according to the research project No. 18-03-00352, using equipment of High Technology Center at BSTU named after V.G. Shukhov.

References

Chernysheva NV, Ageeva MS, Elyan I, Drebezgova MYu (2013) The effect of mineral additives of different genesis on the microstructure of the gypsum-cement stone. Bull. BSTU named after V.G. Shukhov 4:12–18

Christ S (2015) Fiber reinforcement during 3D printing. Mater Lett 139:165–168

De Schutter G, Lesage K, Mechtcherine V, Nerella VN, Habert G, Agusti-Juan, I (2018) Vision of 3D printing with concrete – technical, economic and environmental potentials. Cement Concrete Res

Elistratkin MYu, Minakova AV, Jamil AN, Kukovitsky VV, Eleyan Issa Jamal Issa (2018) Composite binders for finishing compositions. Constr Mater Products 1(2):37–44

Kuprina AA, Lesovik VS, Zagorodnyk LH, Elistratkin MY (2014) Anisotropy of materials properties of natural and man-triggered origin. Res J Appl Sci 9(11):816–819

Lesovik VS, Chulkova IL, Zagordnyuk LK, Volodchenko AA, Yurievich PD (2014) The role of the law of affinity structures in the construction material science by performance of the restoration works. Res J Appl Sci 9(12):1100–1105

Lesovik VS, Popov DYu, Glagolev ES (2017) Textile-concrete is an effective reinforced composite of the future. Constr Mater 3:81–84

Secrieru E, Fataei S, Schröfl C, Mechtcherine V (2017) Study on concrete pumpability combining different laboratory tools and linkage to rheology. Constr Build Mater 144:451–461

Sumskoy DA, Zagorodnyuk LKh, Zhernovskiy IV (2018) Features of the formation of crystalline neoplasms in astringent compositions depending on the technology of their preparation. Bull BSTU named after V.G. Shukhov 6:71–78

Yi WDT, Biranchi P, Suvash CP, Nisar ANM, Ming JT, Kah FL (2017) 3D printing trends in building and construction industry: a review. Virtual Phys Prototyping 12(3):261–276

Zharikov IS, Laketich A, Laketich N (2018) Impact of concrete quality works on concrete strength of monolithic constructions. Constr Mater Products 1(1):51–58

17

Optimization of Formulations of Cement Composites Modified by Calcined Clay Raw Material for Energy Efficient Building Constructions

A. Balykov(✉), T. Nizina, V. Volodin, and D. Korovkin

Department of Building Structures, Ogarev Mordovia State University, Saransk, Russia
artbalrun@yandex.ru

Abstract. The paper presents the results of the study of the influence of formulation and process parameters of dehydrated raw material preparation based on polymineral clay rocks of the Republic of Mordovia used as independent mineral additives to cement composites. The possibility of increasing the studied physical and mechanical parameters of composites by optimizing the mode of clay raw material calcination and the content of the developed modifier is shown.

Keywords: Cement composite · Dehydration · Mineral additive · Clay

1 Introduction

Currently, Portland cement is the main binder in the construction industry. Introduction of fine-grained mineral additives of natural and man-made origin to Portland cement in order to improve the indicators of its physical and mechanical properties and partially replace clinker is one of the effective ways to ensure sustainable development in terms of resource conservation. In recent years, such mineral additives as microsilica and metakaolin have been increasingly used for more rational use of Portland cement and ensure the required level of cement composites characteristics. These modifiers help to increase the density of cement stone by controlling its phase composition and porosity, thereby allowing improvement of physical, mechanical and operational properties of cement composites at reduced cement consumption (Kirsanova et al. 2015; Nizina et al. 2017; Dvorkin et al. 2015).

However, the resources of the above additives do not meet the increasing needs of the construction industry. In this regard, researchers face the challenge of expanding the resource base for the production of mineral additives using available natural raw materials. One of the most promising in this respect are calcined clay rocks (Schulze et al. 2015). At the same time, according to the studies (Rakhimov et al. 2017; Fernandez et al. 2011) results, it was found that kaolinite, montmorillonite and muscovite/illite clays have the highest pozzolanic activity after heat treatment.

The territory of Russia is rich in various types of clays. Ordinary (low-melting) clay in Russia is produced almost everywhere. For example, in the territory of the Republic of Mordovia there are more than fifty deposits of clay rocks, which allows classifying the development of active mineral additives based on clay raw materials as promising task of the construction industry, the solution of which a number of economic, technological and environmental problems of the cement industry both in the region and in country as a whole.

2 Methods and Approaches

Clay from Staroshaygovsky deposit (The Republic of Mordovia) was selected as a raw material for mineral additive development. To carry out the experimental studies, a plan was prepared, which includes 15 experiments allowing variation of the temperature and duration of calcination at three levels (400, 600 and 800 °C and 2, 3 and 4 h, respectively), and the content of mineral additive based on thermally-activated clay in the composition of cement composites on five levels – 2, 6, 10, 14 and 18% of the weight of Portland cement. Also, the additive-free composition (No. 16) was studied in addition to the 15 formulations included in the main block of the experiment plan. Manufacture of cement compositions was carried out at a fixed water-solid ratio of 0.3. The calcined clays were ground in a planetary mill for 1 h. The resulting fine powder was introduced into the cement binder based on Portland cement CEM I 42.5 N produced by Serebryakovcement PJSC. According to the results of the study, optimization of the modified cement binders was carried out and the most effective calcination modes were determined. Rational compositions were determined according to the analysis of an experimental statistical model describing the change in compression resistance of cement composites based on modified calcined clay raw materials:

$$\begin{gathered} y = 67,29 + 3,23 \cdot x_1 + 0,18 \cdot x_2 - 3,99 \cdot x_3 + 1,36 \cdot x_1 \cdot x_2 \\ -0,81 \cdot x_1 \cdot x_3 - 1,38 \cdot x_2 \cdot x_3 + 0,31 \cdot x_1 \cdot x_2 \cdot x_3 - 7,55 \cdot x_1^2 \\ -4,35 \cdot x_2^2 + 4,48 \cdot x_3^2 - 0,91 \cdot x_1^2 \cdot x_2 - 0,56 \cdot x_1 \cdot x_2^2 \\ -0,49 \cdot x_1^2 \cdot x_3 - 2,19 \cdot (x_1 \cdot x_2 \cdot x_3)^2 \end{gathered} \tag{1}$$

Identification of compromise solutions optimal areas for each factor separately was carried out using frequency ranges, which is one of the most descriptive ways to graphically represent the random variable probability density (Lyashenko et al. 2017).

3 Results and Discussion

According to the results of the conducted study, it was determined that a number of modified cement composites can achieve compression resistance equal to 70 ÷ 80 MPa, which is comparable with the control composition No. 16 (Fig. 1). The highest strength characteristics were achieved in compositions 2, 4, 6 and 13 at a content of calcined clay from 2 to 6% of the cement weight.

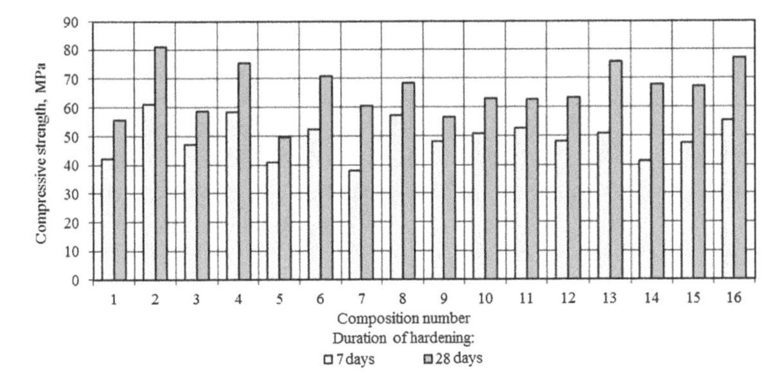

Fig. 1. Compression resistance of modified cement composites at the age of 7 and 28 days

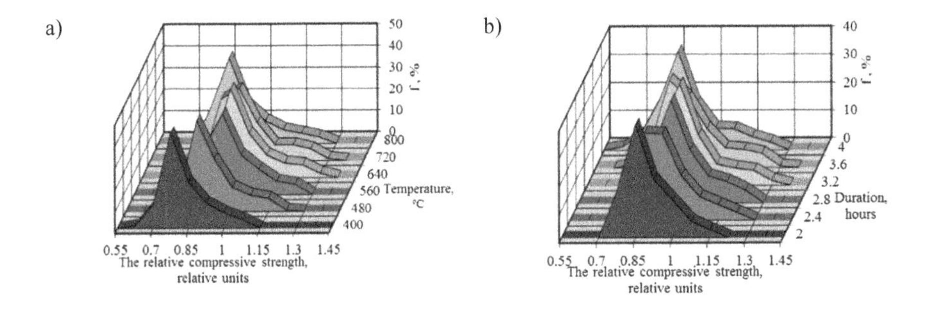

Fig. 2. The range of distribution of the compression resistance of modified cement composites at the age of 28 days: a – from calcination temperature, b – from calcination duration

The analysis of the ES model (1) based on frequency ranges (Fig. 2) showed that the compression resistance corresponding to the control composition can be provided for cement composites with a mineral additive at any studied temperature level and duration of calcination of the clay raw material. At the same time, for the accepted temperature and time intervals of the mineral additive calcination, the total proportion of compositions with enhanced or corresponding to the control composite characteristics varies from 22 to 41% depending on the duration and from 11 to 45% depending on the temperature of calcination. It was found that an increase in clay calcination time from 2 to 3 ÷ 4 h leads to an expansion of the relative values range of modified cement composites strength characteristics from 77.5 ÷ 115 to 62.5 ÷ 130%. Increasing the temperature of calcined clay rocks calcination from 400 to 720 °C allows changing the limit (achievable) range of compression resistance from 62.5 ÷ 107.5 to 85 ÷ 130%, a further increase in temperature leads to a certain decrease in the boundary values of the relative strength indicator to 77.5 (lower boundary) and 122.5% (upper boundary), respectively.

4 Conclusions

According to the results of the study, optimal formulation and process principles for the production of mineral additive based on clay raw materials were determined, which allow increasing compression resistance of modified cement composites in comparison with the additive-free composition. The most effective additives were obtained at calcination time from to 3 to 3.6 h at clay calcination temperature 640 ÷ 720 °C.

The data obtained indicate the prospects and relevance of the development of concrete with modifying additives based on thermally-activated polymineral clays, which allows expanding the range of modified cement composites produced today due to better use of local mineral resources base.

Acknowledgements. The reported study was funded by RFBR and Government of the Republic of Mordovia according to the research project № 18-43-130008.

References

Dvorkin LI, Zhitkovsky VV, Dvorkin OL, Razumovsky AR (2015) Metakaolin is effective mineral additive for concrete. Concrete Technol 9–10(110–111):21–24

Fernandez R, Martizena F, Scrivener KL (2011) The origin of the pozzolanic activity of calcined clay minerals: a comparison between kaolinite, illite and montmorrilonite. Cement Concrete Res 41:113–122

Kirsanova AA, Ionov YV, Orlova AA, Kramar LY (2015) Features of hydration and hardening of cement concretes with additives modifiers containing metakaolin. Cement Appl 2:130–135

Lyashenko TV, Voznesensky VA (2017) Composition-process fields methodology in computational building materials science. Astroprint, Odessa

Nizina TA, Balykov AS, Volodin VV, Korovkin DI (2017) Fiber fine-grained concretes with polyfunctional modifying additives. Mag Civil Eng 4:73–83. https://doi.org/10.18720/MCE.72.9

Rakhimov RZ, Rakhimova NR, Gaifullin AR, Morozov VP (2017) Properties of Portland cement paste incorporated with loamy clay. Geosyst Eng 6:318–325

Schulze SE, Pierkes R, Rickert J (2015) Optimization of cements with calcined clays as supplementary cementations materials. In: Proceedings of XIV international congress on the chemistry of cement, Beijing, China

18

Features of Production of Fine Concretes Based on Clinkerless Binders of Alkaline Mixing

M. Salamanova[1,2(✉)], S.-A. Murtazaev[1,2,3], A. Alashanov[1], and Z. Ismailova[1]

[1] Grozny State Oil Technical University named after Academician M.D. Millionshchikov, Grozny, Russia
Madina_salamanova@mail.ru
[2] Complex Research Institute named after H.I. Ibragimov, Russian Academy of Sciences, Grozny, Russia
[3] Academy of Sciences of the Chechen Republic, Grozny, Russia

Abstract. The paper showed relevance and potential for the development of clinkerless alkaline activation, since at the present the production of currently leading "constructional" binder, Portland cement, has been increasing, and carbon dioxide released during cement production has a negative effect on the ecological situation of separate countries and the whole world. The world community has long been concerned about the problem of switching to clinkerless binders and building composites to replace resource-intensive cement, at least in those areas of construction that do not need its high technical and functional properties. We gave formulations and properties of clinkerless alkaline activation based on highly dispersed mineral components, effective compositions of fine concretes based on the use of the proposed clinkerless alkaline activation cements were obtained. It was theoretically substantiated and practically proved that Brønsted acid sites on the surface of highly active powders accelerated synthesizing silica gel, supported polymerization of silicon-oxygen anions, enhanced ion exchange reactions and stabilized intergranular contact formation.

Keywords: Portland cement · Clinkerless binders · Mineral powders · Alkaline mixer · Liquid glass · High dispersion

1 Introduction

The environmental problem, created by the cement industry, is associated with consumption of large volumes of raw materials and release of huge amounts of carbon dioxide and dust during production of mineral binders into the atmosphere. A promising direction for solving this problem is the use of alkaline activation binders, which can be produced both from wastes of the fuel and energy industry, if present in the region, and using highly dispersed aluminosilicate additives, which chemical composition is characterized by an increased content of aluminum and silicon (Nikiforov et al. 2011; Salamanova et al. 2015; Strokova et al. 2013).

2 Methods and Materials

This paper presents results of researches on the formulation and study of the properties of alkaline mixing binders using of sedimentary and magmatic rocks: quartz sand, limestone, volcanic tuff and silicified marl.

3 Results and Discussion

To prepare highly dispersed powders from the studied rocks, they were preliminarily crushed in a jaw crusher, and then subjected to fine grinding for 1 h in a vibratory ball mill.

At the next stage, a number of Bronsted active crystallization centers on the surface of the mineral powder was studied by the method (Strokova et al. 2013) of determining the exchange capacity with respect to calcium ions (Table 1).

Table 1. Surface activity of fine powders

No.	Mineral powder	Coefficient of activity Ka,%	Coefficient of hydraulic activity G_{HD}	Number of active centers of crystallization, mg eq/g	Specific surface of powders, m^2/kg
1	Quartz sand (QS)	22	1,76	21	810
2	Volcanic Tuff (VT)	37	1,90	34	905
3	Limestone (L)	8	1,24	12	1060
4	Thermo-activated marl (TM) at 700 °C	62	2,03	42	1150
5	Sandstone (S)	10	1,44	16	1020

The coefficient of activity Ka,% The coefficient of hydraulic activity GAMD The number of active centers of crystallization, mg eq/g Specific surface of powders, m^2/kg.

Thus, analyzing the results obtained we can state that the activity coefficients, including the number of active crystallization centers, largely depend on the degree of disclosure of defects formed during their grinding, in combination, all this contributes to an increase in the reactivity of the powders used in concrete as highly dispersed additives.

After confirming the reactivity of the proposed powders, samples of 10 cm in size were prepared using the mixture: a highly dispersed component, fractionated sand, obtained by mixing in a ratio of 55:45% of screening crushing of rocks of the Argun field and fine sand of the Chervlensky field. The mixing was carried out with liquid glass, sodium hydroxide and accelerator of the precipitation of silica gel with sodium fluoride in specified proportions. The prepared samples were hard in normal conditions at a temperature of 20 ± 2 °C, but after 2 days the samples were placed in an oven at a temperature of 40–50 °C for several days. The test results of the investigated fine-grained concretes based on alkaline activation binders are shown in Table 2.

Table 2. Properties of fine concrete based on clinkerless binders of alkaline activation

No compound	Materialconsumption per 1 m^3					Density of concrete, kg/m^3	Strength at compaction, MPa at the age, days	
	Fine powders	Fractionated sand	Na_2SiO_3	NaOH	Na_2SiF_6		7	28
1	QP - 480	1700	72,0	6,0	42,0	2240	11,8	24,7
2	TM - 480	1700	72,0	6,6	41,4	2250	31,4	40,5
3	VT - 480	1700	84,0	6,0	30,0	2246	25,3	34,2
4	L - 480	1700	60,0	3,6	56,4	2235	5,9	14,1
5	S - 480	1700	60,0	3,8	56,2	2239	6,6	15,9

Note: QP - quartz powder; TM - thermo-activated marl at 7000 °C; VT - volcanic tuff; L - limestone flour; S - sandstone; Na_2SiO_3 - sodium liquid glass; Na_2SiF_6 - sodium silicofluoride; NaOH - sodium hydroxide.

The alkaline activation binder with the use of fine powders of thermally activated marl showed the best results, so these particular samples of concrete were examined by Quanta 3D scanning electron microscope (Fig. 1).

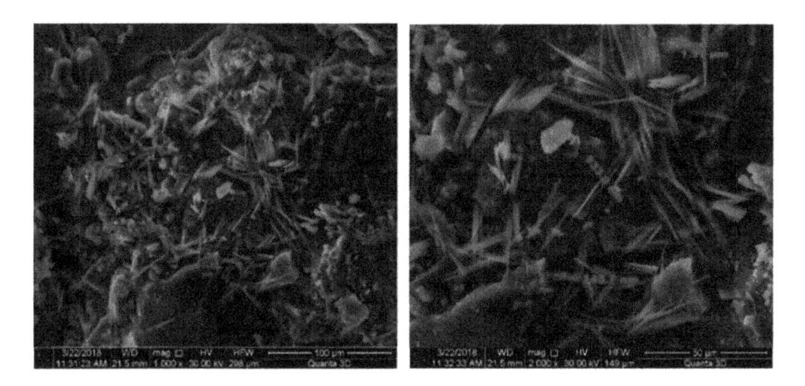

Fig. 1. Micrographs of concrete on thermo-activated alkali activated marl

In the contact area we determined a fairly strong accretion of particles of binder and quartz sand, no defects on the surface in the form of growths or cracks, irregularities of various shapes and sizes, and that individual particles have a needle-fibrous, vitreous structure, which indicated an increased binder activity.

High strength results of fine concrete using clinkerless alkaline mixing on the basis of thermally activated marl are explained by formation of a durable geopolymer stone, represented by a frame aluminosilicate shutter alkaline medium with the formation of a 3D aluminosilicate hydrogel (Murtazaev and Salamanova 2018; Soldatov et al. 2016).

4 Conclusions

Thus, we theoretically justified and practically proved that Brønsted acid sites on the surface of highly active powders accelerated process of synthesizing silica gel, promoted the polymerization of silicon-oxygen anions, enhanced ion exchange reactions and stabilized intergranular contact formation.

The results of the researches significantly expanded the field of application of clinkerless binders on a liquid-glass binder and might enable partial replacement of expensive and energy-intensive portland cement in the construction industry.

References

Murtazaev S-A, Salamanova M (2018) Prospects for the use of thermoactivated raw materials of aluminosilicate nature. Volga Sci J 46(2):65–70

Nikiforov EA, Loganina VI, Simonov EE (2011) The influence of alkaline activation on the structure and properties of diatomite. Bull BSTU Named After V.G. Shukhov (2):30–32

Salamanova MSh, Saidumov MS, Murtazaeva TS-A, Khubaev MS-M (2015) High-quality modified concretes based on mineral additives and superplasticizers of various nature. Sci Anal Mag "Innov Invest" (8):159–163

Soldatov AA, Sariev IV, Zharov MA, Abduraimova MA (2016) Building materials based on liquid glass. In: The collection: actual problems of construction, transport, engineering and technosphere safety materials of the IV-th annual scientific-practical conference of the North-Caucasian Federal University. N.I. Stoyanov (executive editor), pp 192–195

Strokova VV, Zhernovskiy IV, Maksakov AV (2013) Express-method for determining the activity of silica raw material for production, granulated nanostructuring aggregate. Constr Mater (1):38–39

19

Structural Transformations of Low-Temperature Quartz During Mechanoactivation

I. Zhernovsky(✉) and V. Strokova

Belgorod State Technological University named after V.G. Shukhov, Belgorod, Russia

zhernovsky.igor@mail.ru

Abstract. The paper presents the study of mechanoactivation impact on crystal structure of α-quartz. The volume of silicon-oxygen tetrahedron SiO_4^{4-} is accepted as the structural parameter depending on the mechanoactivation degree. The paper compares the dependence of this parameter on temperature, pressure and time of mechanoactivation of α-quartz in a planetary mill.

Keywords: Quartz · Crystal structure · Silicon-oxygen tetrahedron · Mechanoactivation

1 Introduction

The mechanoactivation dispergation of quartz materials is a widespread method of technological processing of this mineral raw material in various fields of technological mineralogy.

In practice of technical petrogenesis of cast stone, forming the basis of synthesis of inorganic silica-containing binding agents (Dmitrieva et al. 2018), the mechanoactivation dispergation of quartz raw material holds a special place.

The result of almost a century-long study of phase and structural transformations of low-temperature quartz during mechanoactivation is the amorphicity of a surface layer of quartz particles and the formation of nanosized β-quartz crystals in α-quartz matrix (Zhernovsky et al. 2018). At the same time there was no study on structural transformations of α-quartz within a quartz particle matrix during mechanoactivation. The only exception is the study by Archipenko et al. (1987, 1990) concerning phase transformations in mechanoactivated α-quartz. The task of this study is to fill the gap in this matter partially.

2 Materials and Methods

The grinding of hydrothermal quartz (Ural) was carried out in the PULVERISETTE 6 classic line planetary mill (Fritsch, Germany) with lining and grinding bodies of tungsten carbide. The milling time made 3, 12, 30, 60, 120, 180, 240, 300 and 360 min.

The diffraction spectra of samples are obtained using ARL 9900 Workstation (λCo). Shooting interval – $2\theta°$:8–80, step angle – 0.02. The specification of structural parameters was carried out in DDM v.1.95e software for the difference curve derivative (Solovyov 2004). 174-ICSD data ($P3_221$) were used as a structural model. Profile function – pseudo-Voigt. The specification of the profile parameters was carried out in the anisotropic approximation. Thermal corrections were specified in anisotropic option. Table 1 shows the experimental results.

Table 1. Coordinates of α-quartz atoms after mechanoactivation

	Milling time, min								
	3	12	30	60	120	180	240	300	360
Si^{*} *x/a*	0.4499	0.463	0.460	0.466	0.465	0.466	0.468	0.469	0.470
O x/a	0.403	0.416	0.413	0.419	0.418	0.419	0.422	0.422	0.423
O y/b	0.2660	0.2702	0.2697	0.2715	0.2668	0.272	0.2697	0.2721	0.2707
O z/c	0.7950	0.7921	0.7919	0.7905	0.7886	0.7868	0.7866	0.7887	0.7876

Note: *Si y/a* = 0, *Si z/c* = 2/3.

3 Results and Discussion

The volume of silicon-oxygen tetrahedrons was chosen as a structural and sensitive quartz parameter on mechanoactivation influence since it depended on several elementary structural parameters: bond length *Si-O*, bond angle *O-Si-O* and twist angle of tetrahedrons (Goryaynov and Ovsyuk 1999).

It is known that mechanoactivation is the result of two processes influencing the material – local thermal influence and impact pressure. Hence, it is advisable to consider structural changes of quartz during mechanoactivation in comparison with changes during thermal and baric impacts. The following were used as references on changes of quartz structural parameters: at a thermal influence – the work of Kihara (1990), at a baric impact – works of Levien et al. (1980), Hazen et al. (1989) and Glinnemann et al. (1992).

The given results show that mechanoactivation alongside with thermal and baric influences transforms the structure of α-quartz by reducing the volume of SiO_4-tetrahedrons. At the same time the dependences of volumes of α-quartz silicon-oxygen tetrahedrons on severity of exposure differ a lot (Figs. 1 and 2).

The energy accumulation by quartz during mechanoactivation, as well as under thermal and baric influence most likely happens due to the reduction of *Si-O* bond length.

The energy accumulated by quartz during mechanoactivation can be estimated by comparing the dependences (Figs. 1a and 2). This comparison confirms that material energy saturation after three-hour mechanoactivation is equivalent to its heating up to 500 °C.

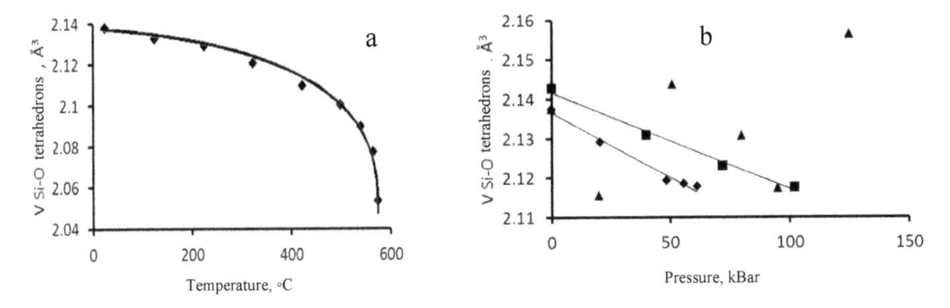

Fig. 1. Calculated dependences of volumes of α-quartz silicon-oxygen tetrahedrons on thermal *(a)* and baric *(b)* influences. *(b)* indicates the following data: ♦– according to Levien et al. (1980), ▲ – according to Hazen et al. (1989), ■ – according to Glinnemann et al. (1992).

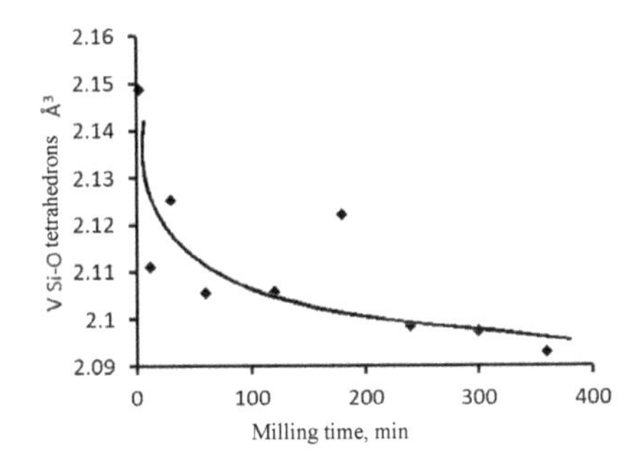

Fig. 2. Dependence of volumes of α-quartz silicon-oxygen tetrahedrons on mechanoactivation time

4 Conclusions

The change of volume of silicon-oxygen tetrahedrons may be considered as an indicator of the α-quartz crystal structure response to the thermal and baric influence, as well as to the mechanoactivation. The difference of these mechanisms of influence is demonstrated by different dependences of this parameter of α-quartz crystal structure on the severity of exposure.

Acknowledgements. The work was performed within the Federal Target Program of Research and Development on Priority Development Fields of Science and Technology Sector in Russia for 2014–2020, unique project number: RFMEFI58317X0063. The authors would like to express gratitude to the Doctor of Geology and Mineralogy, Prof. N.V. Zubkova (MSU) for meaningful consultations.

References

Archipenko DK, Bokyi GB, Grigorieva TN, Koroleva SM, Yusupov TS (1987) On a new quartz phase stable at room temperature and found during tribo-processing (via x-ray diffraction).

Reports of the USSR Academy of Sciences, vol 296, pp 1370–1374
Archipenko DK, Bokyi GB, Grigorieva TN, Koroleva SM, Yusupov TS, Shebanin AP (1990) Deformed quartz structures obtained after mechanoactivation. Reports of the USSR Academy of Sciences, vol 310, pp 874–877
Dmitrieva TV, Strokova VV, Bezrodnykh AA (2018) Influence of the genetic features of soils on the properties of soil-concretes on their basis. Constr Mater Prod 1:69–77
Glinnemann J, King HE, Schulz H, Hahn T, La Placa SJ, Dacol F (1992) Crystal structures of the low-temperature quartz-type phases of SiO2 and GeO2 at elevated pressure. Zeitschrift fur Kristallographie 198:177–212
Goryaynov SV, Ovsyuk NN (1999) Twisting of α-quartz tetrahedrons at pressure close to transition to an amorphous state. J Exp Theor Phys 69:431–435
Hazen RM, Finger LW, Hemley RJ, Mao HK (1989) High-pressure crystal chemistry and amorphization of alpha-quartz. Solid State Commun 72:507–511
Kihara K (1990) An X-ray study of the temperature dependence of the quartz structure. Eur J Mineral 2:63–77
Levien L, Prewitt CT, Weidner DJ (1980) Structure and elastic properties of quartz at pressure. Am Miner 65:920–930
Solovyov LA (2004) Full-profile refinement by derivative difference minimization. J Appl Crystallogr 37:743–749
Zhernovsky IV, Kozhukhova NI, Lebedev MS (2018) Crystallochemical aspects of technological typomorphism of quartz geomaterials during mechanoactivation. In: The collection of papers: fundamental and applied aspects of technological mineralogy. Under the editorship of Doctor of Geology and Mineralogy V.V. Shchiptsov. Karelian Research Center of the RAS, Petrozavodsk, pp 97–100

20

The Law of Similarity and Designing High-Performance Composites

A. Tolstoy(✉), V. Lesovik, E. Glagolev, and L. Zagorodniuk

Belgorod State Technological University named after V.G. Shukhov, Belgorod, Russia
tad56@mail.ru

Abstract. The increasing requirements for the quality of building products and constructions condition the importance of developing new ways of controlling the processes of shaping the structure of composite materials. This paper considers the aspects of dealing with the tasks of designing high-performance powder composites factoring in the law of similarity.

Keywords: Law of similarity · High-performance materials · Powder concrete · Technogenic raw materials

1 Introduction

In our days, erecting special-purpose buildings and structures with complex design, as well as the so-called unique buildings, calls for using high-performance powder concrete, which composition differs from the composition of traditional normal concrete in the increased proportion of cement, higher fineness of grain, complex composition and increased dispersity of aggregate (Bazhenov et al. 2007; Tolstoy et al. 2018a, b; Lesovik et al. 2015). The role of each of the source materials, as well as that of the mechanism of interaction between them, increases manifold, with fundamental concepts of the law of similarity can explain the nature of the processes that occur.

The theoretical foundation for designing high-quality composites is a new cross-disciplinary scientific school: geonics (geomimetics), which employs the results of studies of natural processes and rocks to create building materials of the future (Lesovik 2014; Elistratkin and Kozhukhova 2018; Dmitrieva et al. 2018). This allowed to develop a system for designing powder concrete using raw materials that were specially prepared through geological processes, i.e. they are genetically activated (Tolstoy et al. 2014).

A separate school is further distinguished: crystal energy science, the science about the modern approaches to estimating the quantitative indicators of the properties of materials and explaining the processes of hydrate formation (Lesovik and Evtushenko 2002).

The criteria of applicability of mineral components should be: abundance, accessibility, cost, constant composition (Fig. 1).

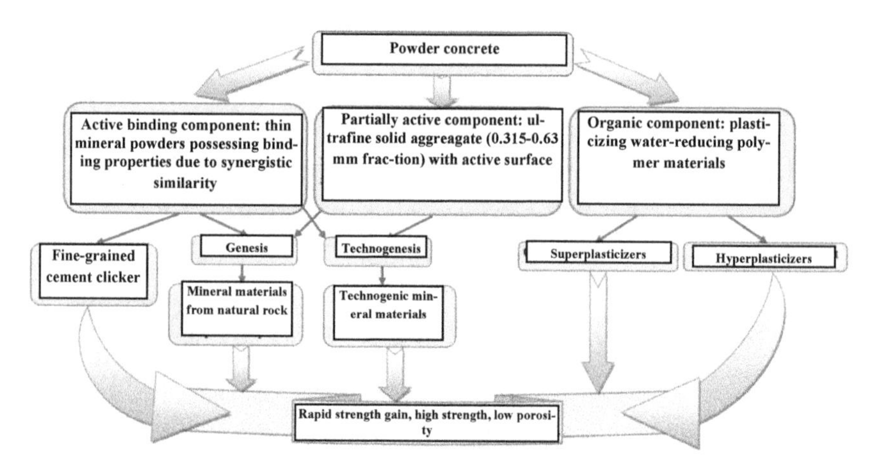

Fig. 1. Types of raw materials for making powder concrete

Here it is necessary to use the "experience" of geological processes, for example, high-performance siltstones that are similar to powder concrete in their strength and other properties. To create a strong and lasting composite of powder concrete, it is necessary to ensure the reliable physical, mechanical and performance characteristics of the material structure taking into account all energy parameters of all the substances involved, as well as the compliance of other properties. This structure should be similar to natural materials in terms of the main properties and genesis.

2 Methods and Approaches

The experiments involved mineral admixtures containing the aluminate and carbonate components, as well as off-the-shelf polymer ones: Melflux 2651, Melment, and finely ground quartzite sandstone, a by-product rock from the Kursk Magnetic Anomaly (KMA).

3 Results and Discussion

At the moment there is no unified approach to the technique of producing high-performance compositions. Some researchers address the issue from the perspective of technological mechanics, other approaches consider the rheological properties of the system, optimization of grain size composition, and the relation between the optimal structure and extreme properties of components (the "law of gauge" by I.A. Rybiev (Rybev 1999)).

The proposed methods of controlling the formation of structure in curing systems of powder mixes with technogenic components allow to achieve the composite strength of 80–100 MPa (Table 1).

Table 1. Comparative figures of normal concrete and powder concrete

Parameter	Values	
	Normal concrete	Powder concrete
Average density, kg/m^3	2200–2500	2300
Compressive strength, MPa	10–50	97.5
Water retention capacity, %	78–80	90
Strength-density ratio	0.17	0.36
Water resistance grade. W	2–4	4
Freeze-thaw resistance grade, F	50–150	300
Wear capacity, kg/m^2	0.7–0.8	0.36
Shrinkage	No fractures	
Thermal conductivity ratio, W/(m K)	0.8–1.2	1.29

A feasible foundation can be represented by the method of estimating the composition of high-density concrete taking into account the law of similarity of properties (Tolstoy et al. 2018a, b).

Dense structure of powder concrete is characterized by a virtually complete absence of pores and microfractures. It was possible to implement this by using the adjusted composition of the curing mass, introducing the necessary number of fine-grained technogenic components, their densest packing, and the self-compacting effect of curing.

Study of the micrographs of the curing compositions revealed the following:

- the microstructure of the cement rock obtained through intergrinding possesses higher homogeneity than that obtained through separate grinding of components;
- growth of needle-shaped crystals that permeate the volume of the material's structure was detected;
- dense neogeneses are present near the aggregate grains;

4 Conclusions

This way, introducing the theory of geonics (geomimetics), particularly the law of similarity, allows to obtain high-performance powder concretes, all components of which possess a high adhesion and similar deformation and thermal characteristics. The first attempt to test this law allowed to obtain composites with the maximum compressive strength of up to 100 MPa.

Acknowledgements. The work has been fulfilled within the project Federal Target Program of Research and Development on "Priority Development Fields of science and technology sector in Russia for 2014–2020", unique project number is RFMEFI58317X0063.

References

Bazhenov IM, Demianova VS, Kalashnikov VI (2007) Modifitcirovannye vysokoprochnye betony [Modified high-performance concretes], ASV, Moscow, 368 p

Dmitrieva TV, Strokova VV, Bezrodnykh AA (2018) Influence of the genetic features of soils on the properties of soil-concretes on their basis. Constr Mater Prod 1(1):69–77

Elistratkin MYu, Kozhukhova MI (2018) Analysis of the factors of increasing the strength of the non-autoclave aerated concrete. Constr Mater Prod 1(1):59–68

Lesovik VS (2014) Geonika (Geomimetika) [Geonics (Geomimetics)]. Primery realizatcii v stroitelnom materialovedenii [Examples of application in building materials science]. Bulletin of BSTU named after V.G. Shoukhov, Belgorod, 206 p

Lesovik VS, Zagorodniuk LK, Chulkova IL, Tolstoy AD, Volodchenko AA (2015) Srodstvo struktur, kak teoreticheskaia osnova proektirovaniia kompozitov budushchego [Similarity of structures as a theoretical foundation of designing the composites of the future]. Stroitelnye materialy [Building materials], no 9, pp 18–22

Lesovik BC, Evtushenko EI (2002) Stabilizatciia svoistv stroitelnykh materialov na osnove tekhnogennogo syria [Stabilizing the properties of building materials on the basis of technogenic raw materials]. Izvestiia VUZov [Bul. HEIs]. 12:40–44

Rybev IA (1999) Otkrytie zakona stvora i vzaimosviaz ego s zakonom kongruentcii v stroitelnom materialovedenii [Discovering the gauging law and its relation to the law of congruence in building materials science]. Stroitelnye materialy [Build Mater] 12:30–31

Tolstoy AD, Lesovik VS, Milkina AS (2018a) Osobennosti struktury betonov novogo pokoleniia s primeneniem tekhnogennykh materialov [Specifics of the structure of new-generation concretes using technogenic materials]. Russ Automob Highw Ind J Sect III Constr Arch. 15(4, 62):588–595

Tolstoy AD, Lesovik VS, Kovaleva IA (2014) Organomineralnye vysokoprochnye dekorativnye kompozitcii [Organic and mineral high-performance decorative compositions]. Bull BSTU Named After V.G. Shoukhov 5:67–69

Tolstoy AD, Lesovik VS, Glagolev ES, Krymova AI (2018b) Synergetics of hardening construction systems. In: IOP conference series: materials science and engineering, vol 327, p 032056. https://doi.org/10.1088/1757-899x/327/3/032056

21

Parameters of Siliciferous Substrate of Photocatalytic Composition Material as a Factor of its Efficiency

Y. Ogurtsova(✉), E. Gubareva, M. Labuzova, and V. Strokova

Department of Materials Science and Technology,
Belgorod State Technological University named after V.G. Shoukhov,
Belgorod, Russia
ogurtsova.y@yandex.ru

Abstract. The article presents the results of the determination pf the properties of the photocatalytic composite material (PCM) of the "TiO_2 – SiO_2" system synthesized by the sol-gel method. The characteristics of siliciferous raw material - diatomite and silica clay, as substrates in the composition of PCM - mineral composition, microstructural features, composition and concentration of active centers on its surface are determined. The dependences of the elemental composition of the surface, the features of the microstructure and photocatalytic activity of PCM on the properties of siliciferous raw material are found. The research shows that the use of diatomite makes it possible to obtain PCM with better characteristics, which is caused by a higher content of the amorphous phase, a more developed and chemically active surface of the particles.

Keywords: Siliciferous raw material · Titanium dioxide · Sol-gel · Photocatalysis · Microstructure · Activity

1 Introduction

The production of photocatalytic composite materials (PCM) of the "TiO_2 – SiO_2" system is aimed at increasing the efficiency of photocatalytic decomposition of pollutants (Arai et al. 2006; Guo et al. 2016).The peculiarities of physical and chemical interaction of siliciferous and titanium-containing components in the synthesis and use of PCM directly affect its photocatalytic activity. In this regard, it is important to study the influence of the properties of siliciferous raw materials on the final characteristics of PCM.

2 Methods and Approaches

As a siliciferous raw material and as a substrate in the composition of the photocatalytic composite material, the Diasil diatomaceous fine dispersed powder (specific surface S_s = 1.39 m^2/g) was used (Diamix, Ulyanovsk region, Russia); fine-ground silica clay (S_s = 1.08 m^2/g) (Alekseevskii deposit, Mordovia, Russia) were used. The determination of the mineral composition of siliciferous raw materials was carried out using an

ARL 9900 WorkStation X-ray fluorescence spectrometer. The peculiarities of microstructure and elemental composition of the surface were studied with the help of TESCAN MIRA 3 LMU high resolution scanning electron microscope. The acid-base characteristics of the surface of siliciferous raw materials were studied using the indicator method (Nechiporenko 2017; Nelyubova et al. 2018).

The production of composite material of the $TiO_2 - SiO_2$ system was obtained by the sol-gel method using a titanium-containing organic precursor—titanium butoxide Ti $(OC_4H_9)_4$ (TBT) (TU 6-09-2738-89, "PROMHIMPERM", Russia). It was dissolved in ethanol, and then the siliciferous substrate (SS) material was introduced into the resulting solution, in a ratio of "TBT/SS" - 4/1. After stirring it on a magnetic mixer, the material was dried and burned at 550 °C.

Then, the tablets were prepared from the obtained materials of the "$TiO_2 - SiO_2$" system. White cements CEM I 52, 5 R (Adana, Turkey) was used as a binder. The ratio $TiO_2 - SiO_2$/cement is 1.3/1. The photocatalytic activity was determined using the photocatalytic decomposition method of the organic pigment Rhodamine B (Rhodamine B, $C_{28}H_{31}ClN_2O_3$). The pigment was applied to the tablets at a concentration of $4 \cdot 10^{-4}$ mol/l. The samples were kept for 4 and 26 h under ultraviolet radiation (UV-A, 1.1 ± 0.1) W/m^2). The evaluation of color change, as an indicator of the effectiveness of self-cleaning of the surface, was carried out according to the Lab color space (coordinate a) using software (Guo et al. 2016).

3 Results and Discussion

Diatomite is a sedimentary biogenic rock consisting of microscopic siliciferous shells of algae (diatoms) with a valve size of 5–200 mcm. The presence of nanoscale pores and elements is shown on the Fig. 1a. Silica clay is of sedimentary biogenic and chemogenic origin, composed mainly of opal-cristobalite silica particles with a size of less than 5 mcm. It is a microporous rock and the content of organic fragments is insignificant (Fig. 1b). The mineral composition of the raw material is similar; a higher content of the amorphous phase is found in the composition of diatomite (Table 1).

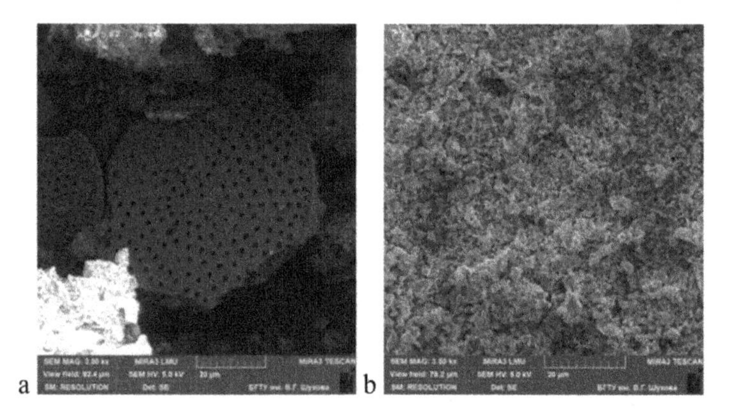

Fig. 1. Microstructure of siliciferous raw materials: a – diatomite, b – silica clay

Table 1. Mineral composition of siliciferous raw material

Siliciferous raw material	*α-Quartz*	*Cristobalite low*	*Tridimite low*	*Illite 2M1*	*Albite*	*Amorphous*
Diatomite	9.35 ± 0.72	40.57 ± 4.22	4.21 ± 0.90	2.44 ± 0.89	0.53 ± 0.20	42.90
Silica clay	17.36 ± 0.72	31.63 ± 1.42	14.63 ± 1.31	3.42 ± 0.65	0.54 ± 0.12	32.42

The presence of a high concentration of acid sites characterized by proton acidity (Brønsted) is noted on the surface of diatomite (Table 2).

Table 2. The composition and concentration of active centers on the surface siliciferous raw materials ($q \cdot 10^{-3}$, mEq/g)

Siliciferous raw material	*pKa*				
	−0.29	+0.80	+7.15	+12.00	+12.80
	Lewis bases	Bronsted acids		Bronsted bases	
Diatomite	53.11	490.43	54.09	47.95	55.80
Silica clay	115.90	No data	20.51	98.68	13.41

The photocatalytic activity of PCM based on diatomite (Table 3) is high and close to the control specimen – the industrial nano-sized Aeroxide TiO_2 P25 photocatalyst.

Table 3. Elimination of Rhodamine B, %

Time	*Aeroxide TiO_2 P25*	*Diatomite*	*TiO_2–SiO_2 (diatomite)*	*Silica clay*	*TiO_2–SiO_2 (Silica clay)*
4 hours	29	9	29	1	5
26 hours	91	61	86	43	68

The analysis of the peculiarities of the microstructure and elemental composition of the surface of synthesized PCM based on diatomite (Fig. 2a) and silica clay (Fig. 2b) shows that the silica particles are partially covered with titanium-containing new formations. The surface of PCM particles based on diatomite is more developed; the distribution of the titanium-containing phase is more even.

Fig. 2. Microstructure of PCM on the basis of: a – diatomite, b – silica clay (with mapping by elements)

4 Conclusions

The siliciferous raw material differs in morphology, concentration of acid-base centers and content of the amorphous phase: the surface of the diatomite is more developed, characterized by a high concentration of proton acid centers; it has a higher content of the amorphous phase in its composition. As a result, the photocatalytic activity of PCM synthesized on the basis of diatomite is higher by 20% in comparison with PCM on the basis of silica clay. In order to improve the efficiency, it is advisable to consider the possibility of pre-activation of silica clay, which will allow using this waste (by-product) rock to produce modern self-cleaning materials.

Acknowledgements. The research was carried out with financial support from Russian Science Foundation grant (project № 19-19-00263).

References

Arai Y, Tanaka K, Khlaifat AL (2006) Photocatalysis of SiO_2-loaded TiO_2. J Mol Catal A: Chem 243:85–88

Guo M-Z, Maury-Ramirez A, Poon CS (2016) Self-cleaning ability of titanium dioxide clear paint coated architectural mortar and its potential in field application. J Clean Prod 112:3583–3588

Nechiporenko AP (2017) Donor-acceptor properties of surface of solid-phase systems. Indicator method, 1st edn. Lan, St. Petersburg

Nelyubova VV, Sivalneva MN, Bondarenko DO, Baskakov PS (2018) Study of activity of polydisperse mineral modifiers via unstandardized techniques. J Phys: Conf Ser 118:012029

22

Multicomponent Binders with Off-Grade Fillers

S.-A. Murtazaev[1,2,3], M. Salamanova[1,2(✉)], M. Saydumov[1], A. Alaskhanov[1], and M. Khubaev[1]

[1] Millionshchikov Grozny State Oil Technical University, Grozny, Russia
Madina_salamanova@mail.ru
[2] Ibragimov Complex Research Institute, RAS, Grozny, Russia
[3] Academy of Sciences of the Chechen Republic, Grozny, Russia

Abstract. The paper deals with issues related to development of multicomponent binders (MCB) and high-quality concretes based on them. The production of such binders is based on the use of finely divided mineral additives of natural and technogenic origin. Particular attention is paid to the aggregate, the strength of coarse aggregate should be at least 20% higher than the strength of concrete, and the maximum particle size should not exceed 8–20 mm. At present, considerable experience was accumulated for production of multicomponent binders, and the results of studies conducted in this direction showed that the raw material potential of the Republic allowed obtaining high-quality class B30-40 concrete, and if we expanded the geography of the use of natural resources by regions of the North Caucasus, we could produce concretes with higher strength.

Keywords: High-quality concretes · Composite binders · Reactive mineral components · Volcanic ash · Thermal power plant (TPP) ash · Fractionated filler

1 Introduction

Concrete is one of the oldest materials, but its potential and possibilities seem inexhaustible (Murtazaev et al. 2016; Nesvetaev et al. 2018; Stelmakh et al. 2018), since at all times of its existence and in the future this material will occupy a leading place among a huge variety of building compositions.

The active component of concrete is cement. It is known that varying finely dispersed mineral additives in its composition results in modern composite materials, which properties will vary in wide ranges (Udodov 2015; Salamanova et al. 2017).

In accordance with GOST 31108-2003, granulated slag, fuel ashes, including acidic or basic fly ash, microsilica, burnt clay, burnt shale, marl, quartz sand, etc. are used as mineral components—main components of cement (Udodov 2015; Murtazaev et al. 2016). Various mineral additives can be used as auxiliary components of cement, which will not significantly increase the water demand of cement and reduce durability of concrete.

2 Methods and Materials

As part of the work carried out in this direction, we developed formulations of multicomponent binders, which include mineral additives of natural and technogenic origin.

The North Caucasus has large reserves of natural raw materials for these developments, the chemical analysis of the mineral components used in the studies is shown in Table 1.

Table 1. The chemical composition of mineral components, wt.%

Type	MgO	Al_2O_3	SiO_2	K_2O	CaO	Fe_2O_3	TiO_2	SO_3	LOI
TPP ash	2,49	23,89	42,88	0,48	4,6	7,95	0,11	0,66	16,9
Volcanic ash	0,20	13,57	73,67	6,00	1,79	1,52	2,85	-	0,40
Limestone flour	0,72	1,55	5,05	0,6	90,14	1,4	-	0,49	-
Quartz powder	6,32	14,99	73,83	1,83	0,6	0,97	1,32	0,14	-

3 Results and Discussion

To produce multicomponent binders, the additives under study were ground in VM-20 laboratory ball vibratory mill for 30 and 40 min. Figure 1 shows dependence of specific surface of mineral additives on the grinding time.

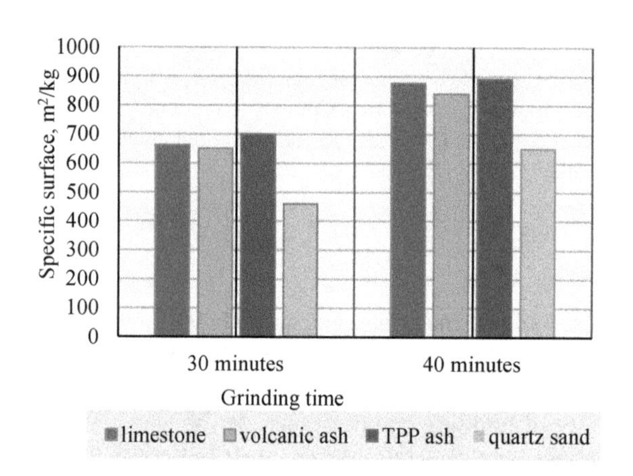

Fig. 1. Specific surface of mineral components

To determine the optimal degree of saturation of Portland cement (PC) – mineral powder (MP) system (PC:MP), samples were prepared from the proposed multicomponent binder formulations and properties (Table 2).

Table 2. Properties of multicomponent binders (MCB)

No.	*Mineral Powder*	*PC:MP*	*Normal density,%*	*Setting time, hour-min*		*Activity, MPa*
				start	*end*	
1	Limestone flour	70/30	25,5	2-05	3-00	35,8
2		60/40	26,8	2-15	3-20	30,4
3	Quartz powder	70/30	24,6	1-30	2-10	41,8
4		60/40	27,0	1-55	2-50	39,7
5	TPP Ash	70/30	26,4	2-10	3-15	34,1
6		60/40	28,1	2-25	3-35	28,2
7	Volcanic ash	70/30	25,2	1-35	2-15	42,6
8		60/40	26,5	2-05	3-00	40,3
9	–	100	25,0	2-20	3-40	48,0

The results of the studies showed that the most rational are the compositions of binders using mineral powders of volcanic ash and quartz powder with a ratio of 70:30%, with a specific surface of 876 m^2/kg and 650 m^2/kg, respectively, with a typical increase in the activity of the binder and a slight increase in normal thickness, and 30% of portland cement are saved.

Next, a concrete mixture with P2 mobility mark was produced, the samples were subjected to heat and humidity treatment (HHT) in a steam chamber at 2 + 3 + 7 + 2 h at an isothermal holding temperature of 80 °C. Table 3 shows the experimental compositions and properties of the studied concretes.

Table 3. The compositions of the studied concretes

No composition	Mineral powder	Consumption of materials, kg/m^3				Average density, kg/m^3	Compressive strength, MPa	
		MCB-70	ACS	DS	B		After HHT	Age 28 days
1	Limestone flour	450	1100	680	220	2430	43,3	38,4
2	Quartz powder	450	1100	680	210	2410	50,2	45,9
3	Volcanic ash	450	1100	680	215	2415	52,1	46,5
4	TPP ash	450	1100	680	230	2420	37,7	35,9
5	PC	450	1100	680	200	2420	51,5	48,6

Note: PC – Portland cement; ACS – Alagir crushed stone fraction 5–20 mm; FS – fractionated fine filled based on the sands of the Alagir and Chervlensk deposits.

We established that the strength of concrete after HHT is 12% higher than the indicators of the strength of concrete after 28 days of natural hardening. The use of MCB-70 with volcanic ash showed the best results on the compressive strength of concrete in comparison with other additives and slightly inferior to similar indicators of control samples (Murtazaev et al. 2016; Stelmakh et al. 2018). The study of operational characteristics (Table 4) showed that the indicators of these properties depend on the composition of the MCB-70 and its activity, as well as on the type and value of the porosity of the material.

Table 4. Operational properties of concrete using MCB-70

Indicators	Mineral powder			
	Limestone flour	Volcanic ash	TPP ash	Quartz powder
MCB-70 activity, MPa	35,8	42,6	34,1	41,8
Compressive strength, MPa	38,4	46,5	35,9	45,9
Flexural strength, MPa	4,1	4,9	3,8	4,4
Porosity, %	9,7	7,6	12,4	6,9
Frost resistance, cycle	F300	F350	F200	F350
Pressure, MPa	1,4	1,8	1,2	1,8
Water absorption, %	4,2	3,5	5,2	3,6
Water resistance, Kr - softening coefficient	0,79	0,89	0,63	0,90

4 Conclusions

Multicomponent binders based on mineral powders of natural and man-made origin allow to obtain high-quality concrete of class of strength B30-40, including for high monolithic construction.

References

Udodov SA (2015) Re-introduction of plasticizer as a tool for controlling the mobility of concrete mix. In: Proceedings of the Kuban State Technological University, no 9, pp 175–185

Murtazaev S-Y, Salamanova MS, Bisultanov RG, Murtazaeva TS-A (2016) High-quality modified concretes using a binder based on a reactive active mineral component. Stroitelnyematerialy, no 8, pp 74–80

Nesvetaev G, Koryanova Y, Zhilnikova T (2018) On effect of superplasticizers and mineral additives on shrinkage of hardened cement paste and concrete. In: MATEC Web of Conferences 27, Cep, 27th R-S-P Seminar, Theoretical Foundation of Civil Engineering (27RSP), TFoCE 2018, p 04018

Stelmakh SA, Nazhuev MP, Shcherban EM, Yanovskaya AV, Cherpakov AV (2018) Selection of the composition for centrifuged concrete, types of centrifuges and compaction modes of concrete mixtures. In: Kim Y-H, Parinov IA, Chang S-H (ed) Physics and Mechanics of New Materials and Their Applications (PHENMA 2018) Abstracts & Schedule, p 337

Salamanova M, Khubaev M, Saidumov M, Murtazaeva T (2017) Self-consolidating concretes with materials of the Chechen Republic and neighboring regions. Int J Environ Sci Educ 11 (18):12719–12724

23

Characterisation of Perovskites in a Calcium Sulfo Aluminate Cement

G. Le Saout[1(✉)], R. Idir[2], and J.-C. Roux[1]

[1] C2MA, IMT Mines Ales, Univ Montpellier, Ales, France
gwenn.le-saout@mines-ales.fr
[2] CEREMA, DIM Project Team, Provins, France

Abstract. Calcium sulfo aluminate cement ($C\overline{S}A$) is a promising low CO_2 footprint alternative to Portland cement. The phase assemblage of a commercial $C\overline{S}A$ cement was investigated by a combination of XRD, SEM-EDX and selective extraction techniques. This study focused on the composition of perovsite phases present in the cement.

Keywords: Calcium sulfo aluminate cement · Perovskite · X-ray diffraction · Scanning electron microscope

1 Introduction

Calcium sulfo aluminate cements ($C\overline{S}A$)[1] were developed by the China Building Material Academy in the seventies. $C\overline{S}A$ have many specific properties compared to Portland cement as fast setting, rapid hardening, shrinkage reduction. This special cement used alone or in combination with calcium sulphates and Portland cement has found applications such as airport runways and roads patching, selfleveling mortars, tile adhesives grouts… (Zhang et al. 1999). This is also a promising low CO2 footprint alternative to Portland cement due to the difference in the amount of energy used to produce $C\overline{S}A$ cements (lower kiln temperatures and energy at the mill to grind). The main raw materials used for making $C\overline{S}A$ cements are bauxite, limestone, clay, and gypsum and this leads to a mineralogical composition very different than Portland cement. While many studies have been carried out on the characterization of Portland cement, few are available concerning $C\overline{S}A$. In this study, we report a characterization of a commercial $C\overline{S}A$ using Rietveld quantitative analysis and scanning electron microscopy.

---

[1] Standard cement chemistry notation is used.
As per this simplified notation: C = CaO, A = Al_2O_3, F = Fe_2O_3, S = SiO_2, $\overline{S}$ = SO_3 and T = TiO_2.

2 Methods and Approaches

The C$\overline{S}$A cement was from a commercial supplier and the chemical characteristics of the cement are given in Table 1.

Table 1. Mineralogical and chemical compositions of the C$\overline{S}$A. Mineralogical composition determined by XRD/Rietveld analysis. Chemical analysis by X-ray fluorescence (DIN 51001)

Minerals	Mass %	Oxides	Mass %
Anhydrite	18.3	SiO_2	8.42
Gypse	2.9	Al_2O_3	19.1
Ye'elimite	31.4	Fe_2O_3	6.94
Belite	21.2	TiO_2	0.76
Perovskite	11.3	K_2O	0.08
Ferrite	5.4	Na_2O	0.02
Merwinite	1.9	CaO	44.9
Calcite	3.5	MgO	1.27
Magnesite	1.5	SO_3	15.2

[a]Loss on ignition measured by calcination until 1025 °C according to ISO 12677

In order to analyze mineralogical composition, X-ray diffraction was performed on cement with a diffractometer BRUKER D8 Advance. Powder samples were analyzed using an incident beam angle (Cu Kα, λ = 1.54 Å) varying between 5 and 70°. Software X'Pert High Score was used to process diffraction patterns and crystals were identified using the Powder Diffraction File database. Rietveld analysis allowed obtaining mass fractions of crystalline phases in the cement.

For the microscopical investigations, powder samples were impregnated using a low viscosity epoxy and polished down to 0.25 μm using diamond pastes. The samples were further coated with carbon (~15 nm) and examined using a Quanta 200 FEG scanning electron microscope (SEM) from FEI coupled to an Oxford Xmax N 80 mm^2 energy dispersive X-ray spectroscopy (EDX) analyser.

In order to improve the characterization of the cement, two different selective dissolution methods were used. In the first method, the silicate phases were removed in a solution of acid salicylic in methanol (Hjorth and Lauren 1971). In addition, a second selective dissolution method was used to get mainly perovskite phases in C$\overline{S}$A by removing ye'elimite, anhydrite, gypsum with 5% Na_2CO_3 solution (Wang 2010). The method was modified to prevent precipitation of $CaCO_3$ by washing the filtered suspension with 6% acetic acid. The filter paper and contents were placed in an oven at 105 °C until a constant weight was reached.

3 Results and Discussion

The main phases observed in the experimental diffraction pattern (Fig. 1a) are the orthorhombic ye'elimite $C_4A_3\overline{S}$ with small amount of the pseudo cubic form, belite β and α'$_{H}$- C_2S and perovskites from the $C\overline{S}A$ clinker. Anhydrite II $C\overline{S}$ is also present as mineral addition. Perovskite family has crystal structures related to the mineral perovskite CT. Ferrite phase $Ca_2(Al_xFe_{1-x})_2O_5$ is usually present in $C\overline{S}A$ and its structure is derived from that of perovskite by the substitution of Al and Fe for Ti, together with ordered omission of oxygen atoms, which causes onehalf of the sheets of octahedral in perovskite to be replaced by chains of tetrahedral (Taylor 1997). To obtain a good Rietveld refinement, it is also necessary to add a perovskite phase CT (Alvarez- Pinazo et al. 2012). The titanium dioxide is present by the use of bauxite, which usually contains some TiO2, as raw materials in the manufacturing process of $C\overline{S}A$ clinker. The peaks associated with this cubic phase are confirmed in the XRD pattern of the $C\overline{S}A$ after the extraction of the main phases (Fig. 1b).

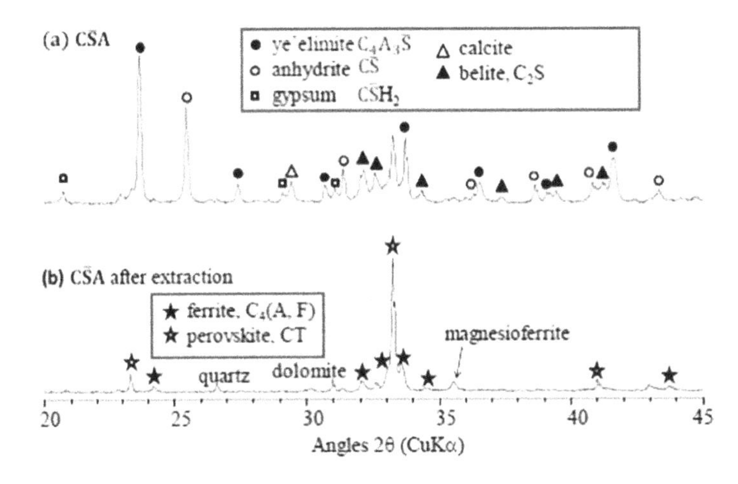

Fig. 1. Diffraction pattern of the cement as received (a), after extraction (b)

However, with the assumed stoichiometry of perovskite CT, the elemental oxide composition TiO_2 calculated from phase content deduced by Rietveld analysis is strongly overestimated in comparison with XRF analysis.

EDX analysis on polished section of $C\overline{S}A$ (Fig. 2a) and $C\overline{S}A$ after extraction revealed an average composition of ferrite $Ca_{1.99}Al_{0.39}Si_{0.1}0Fe_{1.35}Ti_{0.08}Mg_{0.07}O_{5.00}$ not far from the brownmillerite series $Ca_2(Fe_{2-x}Al_x)O_5$. However, the average composition of perovskite $Ca_{2.00}Al_{0.30}Si_{0.22}Fe_{1.06}Ti_{0.40}Mg_{0.05}O_{5.31}$ is very different to the $Ca_2Ti_2O_4$ composition. These compositions were similar to those observed in the ferrite/perovskite phases of a calcium aluminate cement (Gloter et al. 2000). We also observed in some grains perovskite lamellae with high amount of Ti on the scale of few micrometers (Fig. 2b).

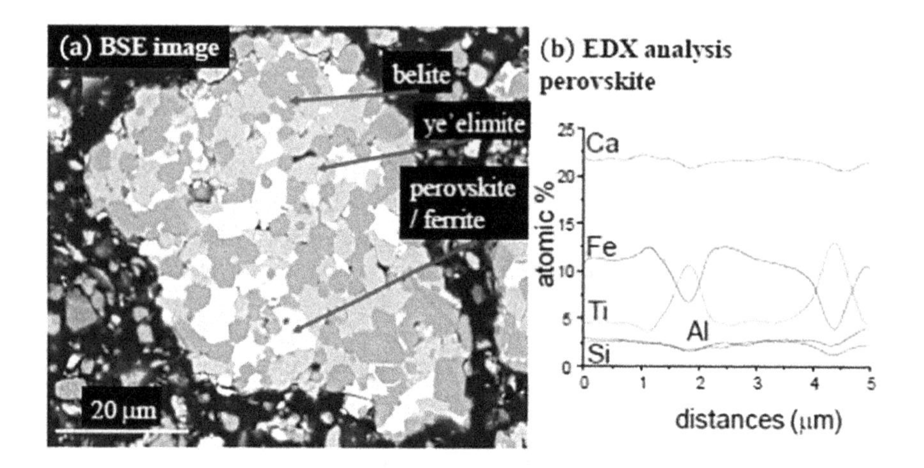

Fig. 2. Back scattering image of the cement (a), example of EDX analysis on perovskite (b)

4 Conclusions

The phase assemblage of a commercial calcium sulfo aluminate cement has been investigated with a special attention to the ferrite- perovskite phases. The ferrite composition is closed to the brownmillerite $Ca_2(Fe_{2-x}Al_x)O_5$ whereas the perovskite shows heterogeneity with important substitution of Ti by Fe, Al and Si.

Acknowledgements. The authors acknowledge A. Diaz (C2MA, IMT Mines Ales) for sample preparation for SEM experiments

References

Alvarez- Pinazo G, Cuesta A, Garcia-Maté M, Santacruz I, Losilla ER, De la Torre AG, Leon-Reina L, Aranda MAG (2012) Rietveld quantitative phase analysis of Yeelimitecontaining cements. Cem Concr Res 42:960–971

Gloter A, Ingrin J, Bouchet D, Scrivener K, Colliex C (2000) TEM evidence of perovskite-brownmillerite coexistence in the $Ca(Al_xFe_{1-x})O_{2.5}$ system with minor amounts of titanium and silicon. Phys. Chem. Miner 27:504–513

Hjorth L, Lauren K-G (1971) Belite in Portland cement. Cem Concr Res 1:27–40

Taylor HFW (1997) Cement Chemistry. Thomas Telford, London

Wang J (2010) Hydration mechanism of cements based on low-CO_2 clinkers containing belite, ye'elimite and calcium alumino ferrite. PhD dissertation, University of Lille I, France

Zhang L, Su M, Wang Y (1999) Development of the use of sulfo- and ferroaluminate cements in China. Adv Cem Res 11:15–21

24

Structuring Features of Mixed Cements on the Basis of Technogenic Products

M. Garkavi[1], A. Artamonov[1], E. Kolodezhnaya[2](✉), A. Pursheva[1], and M. Akhmetzyanova[1]

[1] JSC "Ural-Omega", Magnitogorsk, Russia
[2] Department of Mining and Ecology, Institute of Comprehensive Exploitation of Mineral Resources, Russian Academy of Sciences, Moscow, Russia
gsm@uralomega.ru

Abstract. The structure of mixed cements with mineral additives of different nature enables to divide them into two types: with single-stage and multi-stage structure formation. Characterization of mixed cements is determined, mainly, by nature of the mineral admixture. The index of multi-stage structure formation of mixed cements has been suggested. This index is the value of instantaneous power of structure formation $W\eta$. It has been shown that in case of the other equal conditions the nature of structure formation is determined by the ratio of the components and by milling method of the mixed cement. As a result of analysis of numerous experimental data it has been detected, that the strength of stone, which is formed during hardening of mixed cements, has an extreme nature depending on the content of the mineral admixture. Analysis of the thermodynamic stability of the structure of the stone, which is formed in the process of hardening of mixed cements, allows to divide mixed cements into three groups and identify field of their rational use.

Keywords: Mixed cement · Structure formation · Instantaneous power of structure formation

1 Introduction

Technogenic products have a low hydraulic activity. Therefore, the main condition is to increase their reaction capacity in making the mixed binders, which is achieved by increasing the degree of fineness and concentration of blemishes.

Milling is an energy intensive process operation. So, the energy reduction is achieved by using modern fineness of grind, of which vertical shaft impacted mill is the most effective.

The feature of VIS mill is their high-power rating (more than 10 kW/kg). This allows using them in mechanoactivation processing when solid materials increase capacity of reaction.

The simultaneous grinding and mechanoactivation in VIS mills provide to diversify types of cements (Garkavi et al. 2013).

2 Research Methods

Portland cement clinker, granulated blast-furnace slag, steelmaking slag and flyash were used for mixed cement production of different mixture ratio.

All components of mixed cements were grinded in VIS mills not only for fineness needed but also for their mechanoactivation. The real size of BET surface area is 10000…10400 sq. m/kg, which is an indication of high defect structure and related concentration of surface specific sites.

3 Results and Discussion

During hardening of mixed binding systems several chemical reactions of hydration are connected with various hydraulic activity of their components. Each chemical reaction of hydration, which occurs in a mixed binding system is interconnected with one process of structure formation and affects its development. The character of structure formation of mixed cements with mineral additives of different nature enables to divide them into two types: with a single-stage or multi-stage structure formation (Garkavy 2005). The nature of the mineral admixture shows that mixed cements belong to one or another type.

The multi-stage structure formation means the cyclicity in occurrence of certain structural conditions of mixed binding system. The number of stages of structure formation is determined by the composition of mixed cement such as coagulation or coagulation-condensation contacts.

Prerequisite for the development of multi-stage structure formation is a significant distinction of the apparent activation energies of hydrate-formation of mixed binding components (Garkavy 2005).

Values of apparent activation energy for components of mixed binding is shown in Table 1.

Table 1. Activation energy of hydration of the components of the mixed cements

Component of mixed cement	Activation energy of hydration for the period of, J/mol		
	Induction	Acceleration	Damping
Portland cement clinker	15703	14610	23736
Flyash	47837	17288	31563
Granulated blast-furnace slag	15668	14703	23898
Steelmaking slag	43675	19953	35595

The contribution of components of the mixed cement to the formation of inter-particle contacts can be estimated by the quantity of instantaneous power of structure formation $W\eta$. This value is the product of the rate constant of structure formation process $k\eta$ and the value of the apparent activation energy of this process $E\eta$, i.e. $W\eta = k\eta * E\eta$ (Garkavy 2005).

By the synchronous development of the processes of hydrate and structure formation, in other words, by the one-stage structure formation the correlation is present:

$$W_{\eta \text{mix}} < W_{\eta \text{cl}} + W_{\eta \text{min.ad}} \tag{1}$$

where $W\eta mix$, $W\eta cl$ and $W\eta min.ad$ – instantaneous power of structure formation of the mixed cement, clinker component and mineral admixture, respectively.

Consequently, by one-stage structure formation both components of the mixed cement are involved in the formation of structure, but the predominant influence has a component, which possesses a high instantaneous power structure formation. In case of breach of synchronicity of these processes, that is, by the multi-stage structure formation, the relation is implemented:

$$W_{\eta \text{mix}} = W_{\eta \text{cl}} + W_{\eta \text{min.ad}} \tag{2}$$

It means that in the structure formation components of the mixed cement are consistently involved, the component with the greater instantaneous power of structure formation contributes the most.

The correlation (2) which takes into account the mass fraction φ of the mineral admixture in the mixed cement has the form:

$$W_{\eta \text{mix}} = (1 - \varphi) \cdot W_{\eta \text{cl}} + \varphi \cdot W_{\eta \text{min.ad}} \tag{3}$$

where φ - the mass fraction of mineral admixture in the mixed cement.

From the formula (3) follows, that ceteris paribus, the nature of structure formation is determined by the ratio of the components of the mixed cement. According to the result of the experimental research, the multi-stage structure formation is typical of the mixed cements based on steelmaking slags ($\varphi > 0.3$) and fly ash ($\varphi > 0.5$) (Garkavi et al. 2010).

Thus, the Eq. (3), along with the ratio of the quantity of activation energy of hydrate formation components of the mixed cement can be an indicator of multi-stage structure formation in the binder systems. It is obvious that in the process of hardening of mixed cements with the diverse character of structure formation, intermediate and final structural states with diverse thermodynamic stability are formed. It predetermines their various physical-mechanical and operational characteristics and creates the prerequisites for the creation of rational compositions of mixed cements which have desired properties.

As a result of numerous experimental data, it has been detected (Garkavi et al. 2010; Garkavi et al. 2013), that the strength of stone which is formed in the process of hardening of mixed binders, has an extreme character which depends on the mass fraction of the mineral admixture in the mixed cement.

The extreme dependences, which are shown in Fig. 1, are approximated by the empirical equation:

$$R_{mix} = R_{cl} \cdot exp(b \cdot \phi + c \cdot \phi^2), \tag{4}$$

where Rmix – activity of the mixed cement Rcl - activity of Portland cement clinker; b, c - constants depending on the nature and dispersion of the mineral admixture; φ - the mass fraction of mineral admixture of mixed cement.

From the correlation (4) follows that the value b/c characterizes the mass fraction of mineral admixture of mixed cement φ cr, in which its activity equal to the activity of the Portland cement with no additives. According to the Eq. (3), in the mixed cement the conditions for the development of multi-stage formation are created when the content of mineral admixture is more than φ cr.

4 Conclusion

In agreement with the nature of hardening of mixed cements, they can be divided into three groups with the following rational fields of application:

Group 1 – mixed cements which contain the mineral admixtures $\varphi \leq \varphi cr$ are characterized by single-stage structure formation, they comply with Eq. (1). The mixed cements of this group are substitutes of cements with no additives.

Group 2 - mixed cements with multi-stage structure formation, which satisfy the relation (3). These cements should be used for the manufacture of concretes, which hardened by the heat treatment, as well as for slow-hardening concretes for massive structures.

Group 3 – cements content a high mass fraction of mineral admixture (80%). These cements form the weak strength structure of hardening with low thermodynamic stability. The cements of this group should be used for the production of concrete and low-grade concretes and solutions.

The offered classification of mixed cements allows to identify their rational compositions, on the basis of the properties of mineral admixture and applications.

References

Garkavy MS (2005) Thermodynamic analysis of structural transformations in the binder systems. MSTU, Magnitogorsk, 243 p

Garkavi MS, Hripacheva IS (2010) Optimization of mixed binders with the use of dump steelmaking slag. Build Mater 2:56

Garkavi MS, Hripacheva IS, Artamonov AV (2013) Centrifugal impact grinding cements. Cement Appl 4:106–109

25

Santa Maria Clays as Ceramic Raw Materials

Â. Cerqueira[1(✉)], C. Sequeira[1], D. Terroso[1], S. Moutinho[1], C. Costa[1], and F. Rocha[1,2]

[1] GeoBioTec, Geosciences Department, University of Aveiro, 3810-193 Aveiro, Portugal
angelamcerqueira@ua.pt

[2] RISCO, Civil Engineering Department, University of Aveiro, 3810-193 Aveiro, Portugal

Abstract. There are evidences and records concerning clay exploitation for pottery in the island of Santa Maria (Azores, Portugal) in the past. Nowadays this activity is almost extinct but this is the only island in the archipelago with abundant residual and sedimentary clay deposits. To evaluate the applicability of clays from this island for modern ceramics, a campaign made in May 2017 allowed collecting twenty samples in several outcrops all over the island. All samples were subsequently analyzed in terms of granulometry, mineralogy, chemical composition and physical properties. Results revealed to be interesting, namely concerning mineralogical composition, where phyllosilicates such as Kaolinite are in high percentages. Granulometry also revealed that most part of the samples is composed by fine grain size particles (<63 µm), which can be a good indicator of the existence of resources in great quantity.

Keywords: Santa Maria Island · Pottery · Physical characterization · Potentialities

1 Introduction

Santa Maria, the oldest and most oriental island from the archipelago, is very weathered, as consequence of intense volcanic activity alternated with sea level alterations and intense erosion episodes. There are residual and sedimentary clayey deposits in several parts of the island, one of which known as "Red Desert", corresponding to Feteiras Formation. In addition to the previous, Almagreira Formation is other important deposit of the island. These two in particular were more studied. The abundance of raw materials, gave this island an ancient tradition concerning exploitation of clays for pottery, being known as the "mother island" of clay. From this island, during centuries, white clays were extracted and exported to provide other islands. Recently a pottery oven was discovery "in the middle" of Vila do Porto and it date from the 18th century.

2 Methods and Approaches

Chemical composition was assessed by X-ray fluorescence, qualitative and semi-quantitative mineralogical analyses were carried out by X-ray diffraction and crystallochemistry analyses were carried on a Scanning Electron Microscope (SEM) Hitachi

SU70 with Energy Dispersive X-Ray Spectroscopy (EDS) Brucker QUANTAX 400. Viscosity was assessed with a Haake Viscotester iQ. Plasticity Index was also computed from Atterberg Limits determination.

3 Results and Conclusions

Results are very positive since most part of samples are rich on phyllosilicates (69% to 98%, being kaolinite the most common), fine-grained, and showing adequate plasticity and viscosity. Geological formations, in particular Feteiras and Almagreira Formations, outcrops on a large part of the island, the residual ones showing always intensive alteration, therefore assuring the existence of good reserves.

26

Matrix Instruments for Calculating Costs of Concrete with Multicomponent Binders

T. Kuladzhi[1(✉)], S.-A. Murtazaev[2,3,4], S. Aliev[2], and M. Hubaev[2]

[1] Lomonosov North (Arctic) Federal University, Arkhangelsk, Russia
kuladzhit@list.ru

[2] Millionshchikov Grozny State Oil Technical University, Grozny, Russia

[3] Ibragimov Complex Research Institute, RAS, Grozny, Russia

[4] Academy of Sciences of the Chechen Republic, Grozny, Russia

Abstract. The paper describes methodology for determining the costs of construction products - concrete with multicomponent binders, containing the matrix formula by Professor M.D. Kargopolov, which is recommended a modern micro-prediction tool for the production efficiency of building materials and products, allowing simultaneous calculations of their costs taking into account all the volumes of material costs: cement, fittling, components of binders, etc., as well as wage costs, depreciation etc.

Keywords: Concretes with multicomponent binders · Solar thermal processing · Micro-prediction tools for production efficiency · Composite building materials · Matrix formula by professor M.D. Kargopolov

1 Introduction

Measures are taken in the building complex of Russia at the state level to improve the efficiency and the system for determining the costs of building products. The estimated cost of construction work includes: direct costs, overhead costs and estimated profit (The Civil Code… 1996; On approval… 2004; Taymaskhanov et al. 2012).

To ensure the effectiveness of the production of innovative products the matrix formula by Professor Kargopolov is recommended, providing transparency and accuracy of calculations of the costs of production, taking into account all the territorial conditions affecting the estimated cost of production (Kuladzhi 2014; Kargopolov 2001).

2 Methods and Materials

D.Sc.(Economics) Kargopolov in (Kargopolov 2001) showed that any economic system of economic entities can be represented as a scheme of interacting objects producing a specific product (as part of X), from which part of the output - W in the studied economic system is used inside the system, and the other - (Y) as the final product is taken outside this system. Therefore, to assess the magnitude of costs and results of the effectiveness of each such product, the entire production process П of an organization can be represented by a general structure.

Matrix formula by Professor M.D. Kargopolov is given by:

$$P = (E - AT) - 1 * DT * C \quad (1)$$

where: $P = \|p_j\|$; $j = \overline{1,n}$ – desired column vector of production (full) cost per unit of production (works, services);

E – single matrix n × n;

$A = \|a_{ij}\|$, $i = \overline{1,n}$, $j = \overline{1,n}$ – matrix n × n of consumption rates of own production resources;

$D = \|d_{ij}\|$, $i \in L \cup R$, $j = \overline{1,n}$ – matrix of consumption rates of the primary resources (L – variable, R – constant),

T – transposition mark for matrices A and D.

$C = \|c_i\|$, $i \in L \cup R$ – column vector of the wholesale and procurement prices of primary resources, and if resources are represented by value indicators in matrix D, then in matrix C these resources, respectively, should be denoted by number – 1 (one).

3 Results and Discussion

To calculate the total cost of production of reinforced concrete floor slabs using helioforms, we used actual data on labor costs and costs of materials for the manufacture of reinforced concrete floor slabs of type 10-60.12 under the conditions of Argun reinforced concrete products and structures (Table 1).

Thus, Table 1 presents matrix D in such a way that it consistently (from simple to complex products – a reinforced concrete slab) reveals indicators of the consumption of materials for:

- production of process water: 1 column;
- steam production: 2 column;
- production of complex binders (dry mix: cement, filler and additives "Bio-NM), made in the scientific laboratory of building department of the Grozny State Oil Technical University named after Academician M.D. Millionshchikov (Kargopolov 2001): 3, 4 columns
- production of components of concrete mixes (dry mix: complex binders - CB and crushing residue): 5–6 columns;
- production of 1 m^3 of concrete products under solar thermal treatment: 7–8 columns.

Table 1. Matrix D of consumption rates of primary resources, incl. purchased, for the production of reinforced concrete products and matrix C of wholesale procurement prices of primary resources

Costs	*Water*	*Steam*	*Binder composition*		*Dry concrete mix*		*Reinforced concrete products with solar thermal treatment*		*№*	*Matrix C*
			KB 100	*KB3 50*	*KB 100*	*KB3 50*	*KB 100*	*KB3 50*		*Price*
Capital investments in helioforms (per 1 m^3 of reinforced concrete products), thousand rubles/m^3	0	0	0	0	0	0	0,012	0,012	*1*	0,012
Cement, thousand rubles/t	0	0	0,5	0,254	0	0	0	0	*2*	5
Crushing residue, thousand rubles/t	0	0	0	0	1,5	1,524	0	0	–	0,25
Filler, thousand rubles/t	0	0	0	0,254	0	0	0	0	*2*	1,5
BIO-NM additive, thousand rubles/t	0	0	0,01	0,00508	0	0	0	0	*4*	22
Fitting, thousand rubles/t	0	0	0	0	0	0	0,065	0,065	*5*	5
Electricity and fuel	0,0124	0,0414	0	0	0	0	0,2794	0,2794	*6*	1
Water, thousand rubles/t	0,02365	0	0	0	0	0	0	0	*7*	1
Wages, thousand rubles	0,01	0,025	0	0	0	0	0,3439	0,3439	*8*	1
Equipment maintenance costs (127.8% of salary), thousand rubles	0,0128	0,032	0	0	0	0	0,4395	0,4395	*9*	1
Shop expenses (25% of salary), thousand rubles	0,0025	0,0064	0	0	0	0	0,086	0,086	*10*	1
* Deduction for social insurance (34% for 2011 *), thousand rubles	0,0034	0,0085	0	0	0	0	0,1169	0,1169	*11*	1
Plant costs (20%), thousand rubles	0,002	0,005	0	0	0	0	0,06878	0,06878	*12*	1
Other, thousand rubles	0,0137	0,05	0	0	0	0	0	0	*13*	1

4 Conclusions

Considering that the balance equation of the Nobel Prize winner in economics V.V. Leontiev is a macro-prediction tool for output of products at the national and world levels, the matrix formula by Professor M.D. Kargopolov should be considered as an instrument for micro-prediction of cost indicators of products of any economic entities - companies, households and other subjects, including products of cluster subjects.

References

Kargopolov MD (2001) Interoperable balances of costs and results of production: theory and practice. Monograph. Arkhangelsk: Publishing House of AGSTU, 182 p

Kuladzhi TV (2014) Methodology for evaluating the effectiveness of design solutions in the construction complex. North (Arctic) Federal University - Arkhangelsk: NAFU Publishing House, 296 p

On approval and implementation of the Methodology for determining the cost of construction products in the territory of the Russian Federation (together with "MDS 81-35.2004 ..."). Resolution of the State Construction Committee of Russia of 05.03.2004 No. 15/1 (as amended on 06.16.2014)

Taymaskhanov KE, Bataev DK-S, Murtazaev S-AY, Saidumov MS (2012) Justification of the economic efficiency of the production of concrete composites based on technogenic raw materials. Questions Econ Law 2:124–128

The Civil Code of the Russian Federation (part two) dated January 26, 1996 No. 14-FZ (as amended on July 29, 2017) (as amended and added, entered into force on December 30, 2018)

27

Production of Bleached Cement

D. Mishin(✉) and S. Kovalev

Department of Technology of Cement and Composite Materials,
Institute of Chemical Technology, Belgorod State Technological University
named after V.G. Shukhov, Belgorod, Russia
mishinda.xtsm@yandex.ru

Abstract. The present paper studies a new method of bleaching that allows extending the range of raw materials for white cement. This paper investigates the separate introduction of mineralizers into raw slurry of CJSC Belgorod Cement plant containing 2.78% of Fe_2O_3. The separate introduction of mineralizers provides bleaching effect and complete consumption of free calcium oxide at the temperature of 1250 °C. The bleaching is caused by the reduced content and changed composition of ferroaluminate phase.

Keywords: Clinker bleaching · White cement · Separate introduction · Na_2CO_3 · CaF_2 · Mineralizers

1 Introduction

Currently, white cement is one of the most demanded building materials due to its wide building and technological properties. The demand for this cement grows with the rate of city development. However, the development of white cement industry is limited by strict requirements to the quality of raw materials. To produce cement of graded whiteness, the content of iron oxide in the clinker should not exceed 0.5% (Zubekhin et al. 2004). The number of stock deposits that can provide such content of iron oxide in clinker is relatively low and will gradually decrease with their depletion.

Thus, the development of new technologies extending the range of raw material sources for white cement by involving components with higher iron content becomes very urgent.

2 Methods and Approaches

Materials and Methods. The raw mixture with high content of Fe_2O_3 was represented by dried slurry of CJSC Belgorod Cement plant (Table 1) with the following module characteristics: LSF = 0.91; n = 2.23; p = 1.29.

The raw mixture was doped by mineralizers Na_2CO_3, CaF_2 and $2C_2S{\cdot}CaF_2$ calculated as ignition basis over 100%: The final mixture contained 3.5% of sodium carbonate as R_2O, 1.5% of calcium fluoride and 8.11% of synthesized $2C_2S{\cdot}CaF_2$ (1.5% if expressed as CaF_2). The mineralizers were introduced separately (Mishin et al. 2016).

Table 1. Chemical composition of slurry of CJSC Belgorod Cement plant [%]

Loss on ignition	SiO_2	Al_2O_3	Fe_2O_3	*CaO*	*MgO*	K_2O	Na_2O	SO_3	*other*
34.8	14.23	3.59	2.78	43.12	0.6	0.4	0.11	0.09	0.37

The effect of mineralizers on the phase composition of the clinker was established using XRD analysis using ARL™ X'TRA Powder Diffractometer (Switzerland).

The completion of clinker formation process was assessed by the content of free calcium oxide. The content of CaO_{free} was determined by ethyl-glycerate method (Boutt and Timashev 1973).

The whiteness grade (brightness coefficient) of the clinker was determined by FB-2 reflection meter using reference polished plate of barium sulphate.

3 Results and Discussion

The implementation of separate introduction of mineralizers leads to bleaching of the samples at the temperature of 1250–1300 °C (Fig. 1(1)). The increase of the burning temperature from 1250 to 1300 °C causes gradual return of clinker color to typical black (Fig. 1(1), (2) and (3)).

Fig. 1. Appearance of clinker samples produced by separate introduction at burning temperature of: (1) 1250 °C; (2) 1275 °C; (3) 1300 °C.

In this temperature interval, complete consumption of free calcium oxide is observed, which indicates the completion of clinker formation processes (Table 2).

Table 2. Characteristics of clinker samples produced by separate introduction of $2C_2S{\cdot}CaF_2$ mineralizer

Burning temperature [°C]	*1250*	*1250*	*1275*	*1300*	*1400**
Cooling method	(water)	(air)	(air)	(air)	(air)
Content of CaO_{free} [%]	0.55	0.5	0.3	0.31	0.10
Brightness coefficient [%]	46	41	37.5	30	31

* common plant clinker

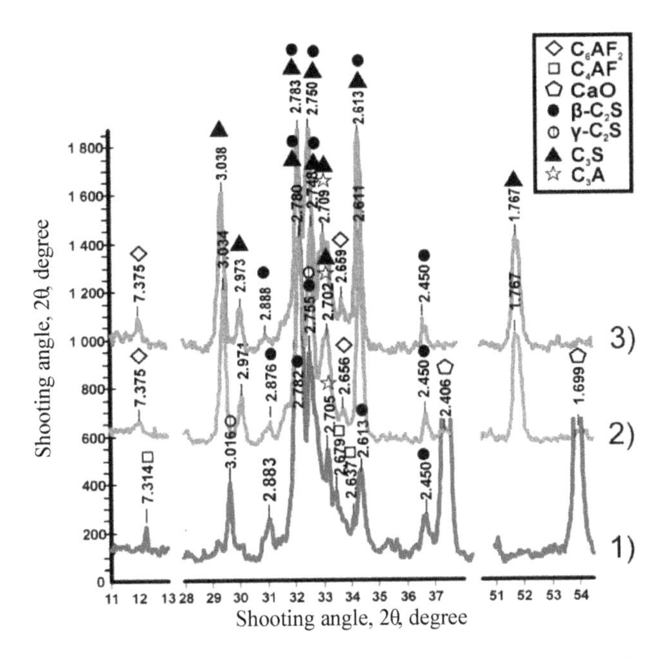

Fig. 2. Phase composition of clinker samples: (1) no additives at 1250 °C; produced by separate introduction at: (2) 1250 °C; (3) 1300 °C.

According to the phase composition analysis, at the burning temperature of 1250 °C (Figs. 2(1), (2)), in the samples produced by separate introduction, the composition of ferroaluminate phase C_4AF (d, Å: 7.314; 2.679; 2.637) shifts into the region of compositions with a higher iron content and approaches that of C_6AF_2 (d, Å: 7.375; 2.656). The intensity of reflections of the ferroaluminate phase decreases, which tells about the decrease in its content in the clinker. Ferroaluminate phase is the most coloring among the clinker components (Zubekhin et al. 2004). In this connection, the decreased amount of ferroaluminate phase leads to increased brightness coefficient of the clinker.

The increased burning temperature up to 1300 °C (Fig. 2(3)) causes increased intensity of ferroaluminate phase reflections (d, Å: 7.375; 2.659), which witnesses its increased content in the clinker. As a result, the brightness coefficient of the clinker reduces from 41% down to 30%.

4 Conclusions

The separate introduction of mineralizer allows bleaching the clinker with the iron oxide content of more than 0.5%. With the iron oxide content in raw mixture of 2.78%, the brightness coefficient increases from 31% up to 46%.

The increase in the brightness coefficient is connected with decreased content of calcium ferroaluminate phases in the clinker: C_4AF, C_6AF_2.

The implementation of separate introduction of mineralizer will allow extending the range of sources for white cement production involving raw components with increased iron content.

The separate introduction of mineralizer in the presence of increased iron oxide content allows achieving complete consumption of free calcium oxide at 1250 °C.

Acknowledgements. The work is realized in the framework of the Program of flagship university development on the base of Belgorod State Technological University named after V.G. Shoukhov, using equipment of High Technology Center at BSTU named after V.G. Shoukhov.

References

Boutt YM, Timashev VV (1973) Practical guide on chemical technologies of binders: guide. Higher school, Moscow, 504 p. (in Russian)

Mishin DA, Kovalev SV, Chekulaev VG (2016) Reason for reduced efficacy of mineralizers for burning Portland cement clinker. Bulletin of BSTU n.a. V.G. Shoukhov, no 5, p 161–166 (in Russian)

Zubekhin AP, Golovanova SP, Kirsanov PV (2004) White Portland cement. Zubekhin AP (ed) J. "Bulletin of HEIs North-Caucasus region", Rostov on Don, 264 p (in Russian)

28

Activation of Cement in a Jet Mill

S. Titov(✉) and A. Kazakov

Federal State Budgetary Educational Institution, Russian University of Transport (MIIT), Moscow, Russia
titovs3094@yandex.ru

Abstract. Mechanical treatment of cement in a jet mill leads to an increase in the strength of cement-sand stone in compression and during bending. The effect of increased activity is achieved by changing the shape of cement particles from angular to rounded.

Keywords: Cement · Jet mill · Cement activity · Compressive strength · Bending strength · Particle shape factor

1 Introduction

Activation is a set of measures aimed at increasing the activity of cement by exposing it to various mechanical methods. As a result of such exposure, physical and chemical processes of different nature occur in the raw materials, leading to a change in the characteristics of the final product (Kuznetsova and Sulimenko 1985). As a rule, during mechanical activation, cement is crushed with an increase in its specific surface.

In a jet mill, activation occurs due to processing of cement in turbulent swirling flows (Korchakov 1986), leading not only to grinding, but also to transformation of the shape of particles from angular to rounded.

2 Methods and Approaches

The analysis of the shape of cement particles modified in a jet mill was performed using the proposed shape criterion:

$$k_i = (S/P^2)^i \tag{1}$$

where S – area of the particle in the plane of its cross section, P – length of the perimeter in the cross section of the particle; i – order of the criterion (i =1/n…,1,…,n, where n – counting numerals).

To determine the degree of influence of the particle shape on the properties of cement systems, standard samples were tested and the results obtained were processed using mathematical statistics methods.

3 Results and Discussion

It was found that processing in a jet type mill leads to an increase in dispersity of cement from 5% to 0.8–1.2% on sieve No. 008 with a non-significant change in normal density of the cement paste.

Increasing the fineness of cement grinding causes, as a rule, an increase in mixing water consumption to obtain a mixture of equal workability. But the experiment showed the opposite – the water-cement ratio of the solution prepared on cement, processed in a jet mill, not only did not increase, but also decreased. This is due to the change in the shape of cement particles from angular to rounded, which was statistically justified using the proposed shape criterion k_i (1).

The results of the tests showed an increase in strength of the samples based on portland cement processed in a jet mill for compression by 38–68% and by bending by 12–25% compared to non-activated cement.

4 Conclusions

The increase in compressive strength and bending of the samples based on cement processed in a jet mill is explained by the change in the shape of its particles from angular to rounded, which is associated with the characteristics of the vortex activation method.

Acknowledgements. The authors express their gratitude to the research supervisor, D.Sc. (Eng), prof. Kondrashchenko V.I.

References

Korchakov VG (1986) Aerodynamics of flows in jet mills when grinding silicate materials, Ph.D. (Eng) thesis, Kharkov, 168 p

Kuznetsova TV, Sulimenko LM (1985) Mechanical activation of portland cement raw mixtures. Cement, no 4, pp 20–21

29

Abnormal Mineral Formation in Aluminate Cement Stone

I. Zhernovsky[(✉)], V. Strokova, V. Nelyubova, Yu. Ogurtsova, and M. Rykunova

Belgorod State Technological University named after V.G. Shukhov, Belgorod, Russia
zhernovsky.igor@mail.ru, zhernovskiy.iv@bstu.ru

Abstract. The paper describes the results of the study on the correlation between the strength properties and the concentration of calcium aluminate hydrates during long hardening of aluminate cement in aqueous medium. It shows the abnormal nature of hydrate mineral formation in the system of hardening aluminate cement.

Keywords: Aluminate cement · Calcium aluminate hydrates

1 Introduction

The properties of any composite materials directly depend on the characteristics of their base matrix. This also fully applies to composites based on binding agents with various composition, which quality indicators are set as a result of system hardening (Strokova et al. 2015; Kozhukhova et al. 2016; Nelyubova et al. 2017; Dmitrieva et al. 2018; Shulpekov et al. 2018). In this regard the study of phase formation in such systems seems quite relevant.

Construction materials based on aluminate cements have different properties than those based on portland cement. In particular, they are characterized by a high thermal stability and resistance to acid corrosion (Kuznetsova 1986; Kuznetsova et al. 1989).

The calcium aluminate hydrates, which are formed through hardening of aluminate cement, represent the following phases: CAH_{10}, C_2AH_8, C_3AH_6, AH_3 (gibbsite). From this point onward the ‘cement’ notation in formulas of chemical compounds will be used (C – CaO, A – Al_2O_3, H – OH).

The hydrate mineral formation in aluminate cements is characterized by the dependence of C-A-H phases on temperature. At a temperature below 15 °C the main reaction-active CA phase is hydrated according to the following scheme: $CA + 10H \rightarrow CAH_{10}$. At room temperature the C-A-H is formed according to the following scheme: $2CA + 11H \rightarrow C_2AH_8 + AH_3$. At higher temperatures (above 28 °C) the metastable hydrates CAH_{10} and C_2AH_8 will transform into stable hydrate C_3AH_6 according to the following equations: $3CAH_{10} \rightarrow C_3AH_6 + 2AH_3 + 18H$ and $3C_2AH_8 \rightarrow 2C_3AH_6 + AH_3 + 9H$ (Rashid et al. 1994).

Thus, according to insights into hydration of aluminate cements, the hydrate phases during long hydration shall be presented as follows: $C_3AH_6 + AH_3$.

The paper provides the results of hydration hardening of aluminate cement with long-term exposure to aqueous medium, which contradict the above ideas.

2 Materials and Methods

The aluminate cement GTS-50 produced by JSC Pashiysky Cement and Metallurgical Plant (Perm Region, Russian Federation) was used in the study as a binding agent. Cubes with a 2 cm side are formed from cement paste (water-cement ratio equal 0.3). After hardening on the 1st day the samples were taken from a mold and placed into the desiccator with 100% air humidity. After 28 days the samples were dried in a drying cabinet at 80 °C and placed in water, from which they were taken for further study in 1, 2, 3 and 4 months.

The study methods included the compression strength test on PGM-100 MG4 press (average against three measurements) and the quantitative full-scale XRF to define potential changes in mineral composition.

The diffraction spectra of samples were obtained via the ARL X'TRA diffractometer (λ_{Cu}) in the range of diffraction angles 2θ° = 4-64, step angle – 0.02°. The diffraction spectra were smoothened prior to treatment.

The quantitative full-scale XRF was carried out via the DDM v.1.95e software for a difference curve derivative (Solovyov 2004).

3 Results and Discussion

According to XRF, the mineral composition of GTS-50 cement is presented by the following crystal formations (wt.%): CA (28.3), CA_2 (9.2), $C_{12}A_7$ (1.4), akermanite-gelenite (15.6), β-C_2S (10.0), α'$_H$-C_2S (10.2), wollastonite 2M (7.6), dolomite (10.6), perovskite (7.2).

X-ray diagnostics of mineral phases in hydrated samples of aluminate cement indicated the presence of CAH_{10}, C_2AH_8, C_3AH_6, AH_3, akermanite-gelenite, β-C_2S, α'$_H$-C_2S, wollastonite 2M, dolomite and perovskite.

Due to lack of C_2AH_8 structural model, the quantitative XRF was combined with the approach suggested by Cuberos et al. (2009).

As data show (Table 1), for nearly 5 months (28 days in damp atmosphere and 4 months in water) CAH_{10} remains the main hydrate phase. The reduction of CAH_{10} concentration on the 2nd month in water, which correlates well with the increase of crystal gibbsite (AH_3), does not lead to the similar increase of C_3AH_6. Most likely it is caused by the formation of C_3AH_6 in its cryptocrystalline state.

The abnormal nature of hydrate mineral formation includes the increase of CAH_{10} metastable hydrate concentration starting from the 3rd month. At the same time the temporal change of C_3AH_6 concentration coincides well with the change of compression strength of the studied samples (Fig. 1).

Table 1. Concentration of hydrate phases (wt. %)

Time in aqueous medium, months	CAH_{10}	C_2AH_8	C_3AH_6	AH_3
0	18.8	3.5	3.1	2.0
1	26.5	4.4	8.2	5.9
2	9.0	2.1	6.8	25.1
3	33.5	3.5	4.7	6.3
4	37.2	5.8	5.6	3.0

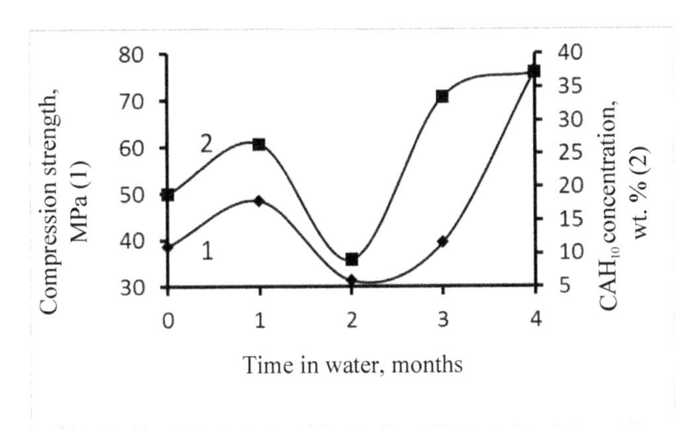

Fig. 1. Compression strength of aluminate cement stone and CAH_{10} concentration depending on time

4 Conclusions

The paper describes the results of the study on the correlation between the strength properties and the concentration of calcium aluminate hydrates during long hardening of aluminate cement in aqueous medium. At present it is impossible to give unambiguous interpretation of the observed abnormal compositional changes of hydrate phases of hydrated aluminate cement during long hardening.

Acknowledgements. The study is carried out in the framework of the State Task of the RF Ministry of Education and Science No. 7.872.2017/4.6. Development of principles for the design of ecologically positive composite materials with prolonged bioresistance 2017–2019.

References

Cuberos AJM, De la Torre ÁG, Martín-Sedeño MC, Moreno-Real L, Merlini M, Ornez LM, Aranda MAG (2009) Phase development in conventional and active belite cement pastes by Rietveld analysis and chemical constraints. Cem Concr Res 39:833–842

Dmitrieva TV, Strokova VV, Bezrodnykh AA (2018) Influence of the genetic features of soils on the properties of soil-concretes on their basis. Constr Mater Prod 1(1):69–77

Kozhukhova NI, Chizhov RV, Zhernovsky IV, Strokova VV (2016) Structure formation of geopolymer perlite binder vs. type of alkali activating agent. ARPN J Eng Appl Sci 11 (20):12275–12281

Kuznetsova TV (1986) Aluminate and sulfo-aluminate cements, Moscow

Kuznetsova TV, Kudryashov IV, Timashev VV (1989) Physical chemistry of binding materials, Moscow

Nelyubova V, Pavlenko N, Netsvet D (2015) Cellular composites with ambient and autoclaved type of hardening with application of nanostructured binder. In: IOP Conference series: materials science and engineering, vol 96, no 1, p 012010

Nelyubova VV, Strokova VV, Sumin AV, Jernovskiy IV (2017) The structure formation of the cellular concrete with nanostructured modifier. Key Eng Mater 729:99–103

Rashid S, Barnes P, Bensted J, Turrillas X (1994) Conversion of calcium aluminate cement hydrates re-examined with synchrotron energy-dispersive diffraction. J Mater Sci Lett 13:1232–1234

Shulpekov AM, Lepakova OK, Radishevskaya NI (2018) Phase – and structural formation in the TIO2-AL-C system in the SHS process. Chem Bull 1(1):4–11

Solovyov LA (2004) Full-profile refinement by derivative difference minimization. J Appl Crystallogr 37:743–749

Strokova VV, Botsman LN, Ogurtsova YN (2015) Impact of epicrystallization modifying on characteristics of cement rock and concrete. Int J Appl Eng Res 10(24):45169–45175

30

High-Quality Concretes for Foundations of the Multifunctional High-Rise Complex (MHC)

S.-A. Murtazaev[1,2,3], M. Saydumov[1(✉)], A. Alaskhanov[1], and M. Nakhaev[4]

[1] Millionshchikov Grozny State Oil Technical University, Grozny, Russia
saidumov_m@mail.ru
[2] Ibragimov Complex Research Institute, RAS, Grozny, Russia
[3] Academy of Sciences of the Chechen Republic, Grozny, Russia
[4] Chechen State University, Grozny, Russia

Abstract. The paper presents results of studies of monolithic concrete mixes and concretes produced with the integrated use of local natural and technogenic raw materials, including waste scrap and crushed bricks. We developed optimal compositions of monolithic concretes and studied their technological and physical-mechanical properties.

Keywords: High-strength concrete · High-quality concrete mix · Filled binder · Technogenic wastes · Mineral technogenic filler · Monolithic concrete

1 Introduction

The modern materials science and construction now deal with an important national economic and engineering problem: development of efficient technologies for producing high-strength monolithic concretes through the integrated use of technogenic raw materials to obtain secondary raw materials for concrete, while eliminating the enormous environmental damage caused by waste "cemeteries" (Bazhenov et al. 2006; Salamanova et al. 2017; Lesovik et al. 2012; Murtazaev et al. 2009; Kaprielov et al. 2018) in Russia and the world and, in particular, for the Chechen Republic, considering construction of 435-m high-rise complex «Akhmat-Tower» in Grozny City (Udodov 2015; Koryanova et al. 2018; Salamanova and Murtazaev 2018).

2 Methods and Materials

Our experimental studies used local additive-free Portland cement of PC 500 D0 grade as a binder. Natural sand from the Chervlensky deposit of the Chechen Republic was used as a fine filler. Local gravel of 5–20 mm fractions from the Argunsky and Sernovodsky deposits of the Chechen Republic and imported crushed stone of 5–20 mm fraction from granite-diabase rocks of the Alagirsky deposit of the Republic of North Ossetia-Alania were used as a coarse filler.

As plastifying agents in accordance with GOST 24211-2008 "Additives for concrete. General technical requirements" modern additives of various manufacturers of building chemicals (POLYPLAST, TOKAR, etc.) were used.

The raw materials for the production of dispersed technogenic mineral fillers (DTMF) were local materials, mainly technogenic, namely concrete scrap, crushed ceramic bricks (CCB), ash and slag mixture from the Grozny heat and power plant (HPP) and very small non-conditioned quartz sands ones were used in comparative tests.

All the DTMFs were ground for 5 min in MV-20-EX laboratory vibratory ball mill with a loading volume of 5–6 L to obtain a specific surface of 450–600 m^2/kg.

3 Results and Discussion

The Filled Binder (FB) formulation was developed and investigated with activity 60–71 MPa with concrete scrap and CCB fillers with ratio 70:30%. The proportion of the mixture of filler in FB was 25 and 40% by weight of the binder.

Due to the fact that for designing the underground part of the Akhmat Tower multifunctional complex, concrete of different strength classes (B40, B75-B80) was laid, the task was to develop high-quality concrete mixes (HCM), starting from the middle B40-B50 classes and ending with high-strength concrete of B80-B90 classes, with the integrated use of local raw materials, including with technogenic nature.

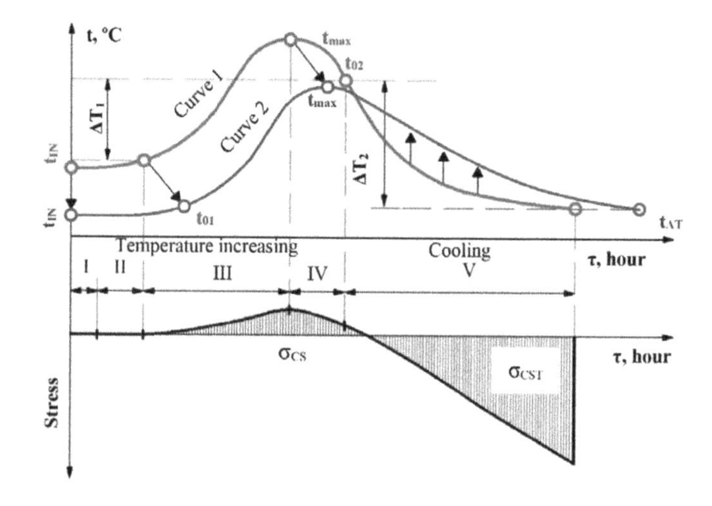

Fig. 1. Temperature change and stress characteristic in fresh concrete with limited deformation: σ_{CS} - internal compressive stresses; σ_{CST} - the same, tensile; t_{IN} - initial temperature of concrete mixture; tmax - the same, maximum; t_{AT} - ambient temperature (air); I, II, ... V - stages (periods) of process of heat dissipation of the concrete mix in time; curve 1 - kinetics of heat dissipation of concrete with PC; curve 2 - the same with FB

In massive building constructions, such as the Akhmat Tower MFC base plate, because of their large dimensions, as a rule, the heat from cement hydration is slowly released into the air or into adjacent structural elements, as a result of which the core of the monolithic element heats up much faster and stronger than the shell. Therefore, we investigated thermophysical processes of the developed compositions.

Figure 1 schematically shows the dependence of temperature and voltage due to external pressure generated by this concreting technology.

The dependence of heat release curves in time is conventionally divided into 5 stages (Table 1).

Table 1. The main stages (periods) of the process of heat dissipation in time of concrete mixes with various binders

Stage №	*Duration, h*		*Description*
	With PC	*With FB*	
1	0–2	0–8	The initial stage without raising the temperature of the concrete mix (dormant period). This period is significantly increased due to the use of surface-active substances (surfactants) in the composition of FB, hardening retarders, etc.
2	2–6	8–15	Temperature increase due to hydration of the binder, no measurable stress, because in the plastic concrete, thermal expansions are converted to relative compression. At the end of this stage, the temperature is referred to as “the first temperature at zero stress” t_{01}
3	6–13	15–24	Further heating of the concrete, the strength of the concrete increases and a compressive stress is formed, partially decreasing because of relaxation. Stage III ends when the maximum temperature tmax is reached
4	13–24	24–72	Heat transfer prevails: the temperature of concrete and compressive stress in concrete decrease, a part of compressive stress decreases because of relaxation. The “second temperature at zero stress” t_{02} is reached, which significantly exceeds t_{01} in cooling rate and age of concrete
5	24–72	72–144	Further cooling and increasing tensile stress, which are partially reduced due to relaxation. If the tensile stress reaches the tensile strength of the concrete under tension (at $\Delta T_{КРИТ}$) through cracks are formed

According to calculations (Kaprielov et al. 2018), the temperature difference between the upper surface layers of concrete slab and outside air ΔT1 should be no more than 20 °C, and the difference between the side layers ΔT2 - no more than 30 °C.

4 Conclusions

Thus, analysis of data confirms the effectiveness of the use of FB with DTMF in high-quality concretes used for concreting massive structures. We established that the peak of the maximum heat release tmax from the exotherm of cement in (massive) concrete on HB was reduced by 30–35% in comparison to the concrete with PC (from 70–75 °C to 50–55 °C).

References

Bazhenov YM, Demyanova BC, Kalashnikov VI (2006) Modified high-quality concretes. Publishing House of the Association of Construction Universities, Moscow, 368 p

Kaprielov SS, Sheinfeld AV, Al-Omais D (2018) Experience in the production and quality control of high-strength concrete on the construction of the high-altitude complex "OKO" in Moscow-City International Business Center. Industrial and Civil Construction, no 1, pp 18–24

Koryanova YI, Rezantsev NE, Shumilova AS (2018) Materials and structures used in the construction of high-rise buildings - from tradition to innovation. Alley Sci 6(4):95–99

Lesovik BC, Murtazaev SAY, Saydumov MS (2012) Construction composites based on screenings of crushing of concrete scrap and rocks. MUP "Typography", Groznyy, 192 p

Murtazaev S-AY, Bataev DK-S, Ismailova ZK (2009) Fine-grained concretes based on fillers from secondary raw materials. Comtechprint, Moscow, 142 p

Salamanova M, Khubaev M, Saidumov M, Murtazayeva T (2017) Int J Environ Sci Educ 11 (18):12719–12724

Salamanova MS, Murtazaev SAY (2018) Clinker-free binders based on finely dispersed mineral components. In: Collection: Ibausil Conference Proceedings, pp 707–714

Udodov SA (2015) Re-introduction of plasticizer as a tool for controlling the mobility of concrete mix. In: Proceedings of the Kuban State Technological University, no 9, pp 175–185

31

Geonics (Geomimetics) as a Theoretical Basis for New Generation Compositing

V. Lesovik[1(✉)], A. Volodchenko[1], E. Glagolev[1], I. Lashina[1], and H.-B. Fischer[2]

[1] Belgorod State Technological University Named After V.G. Shukhov, Belgorod, Russia
naukavs@mail.ru
[2] Bauhaus-Universität Weimar, Weimar, Germany

Abstract. The article introduces basic principles of a new transdisciplinary research area geonics (geomimetics) in the construction material science. This research area differs from bionics, which uses knowledge about nature to solve engineering problems. The purpose of geonics (geomimetics) is to solve engineering problems with the account of knowledge about geologic and cosmochemical processes.

Keywords: Geonics · Anthropogenic methasomatism in the construction material science · The law of consanguinity · Construction composites

1 Introduction

Nowadays many scientists are developing construction composites of new generations (Delgado et al. 2015; Ahadi 2011). Basing on the theoretical background of geonics (geomimetics) it is possible to reduce power consuming in construction materials production by application of energy of geological and cosmochemical processes, energy efficient raw material, specially treated with geological and cosmochemical processes.

So, production technologies for a wide range of composite binding materials (including water resistant and freeze proof gypsum binders) with the application of new raw material (Lesovik et al. 2014; Zagorodnyuk et al. 2018) with high free internal power have been suggested. Clay rocks with an incomplete stage of mineral formation and a sediment genesis zone, effusive rocks with an amorphous and cryptocrystalline structure, quartz rocks of greenschist coal ranging with crystalline defects and inclusions of mineral formation, gas and air inclusions, some kinds of anthropogenic materials and others refer to these materials (Elistratkin and Kozhukhova 2018).

2 Methods and Approaches

The law of consanguinity has been formulated within the theoretical backgrounds of geonics (geomimetics). It implies designing layered composites and maintenance systems at nano-, micro-, and macro- levels similar to basic matrix that enhances significantly material adhesion and durability. The implementation of this law allows creating a composite which components have close deformation and temperature characteristics.

The theory of anthropogenic methasomatosis in the construction material science has been suggested. This stage of composite materials evolution is characterized by composite ability to adjust to changing conditions during building and construction service. Designing construction composites with the account of anthropogenic methasomatosis theory in the construction material science provides possibility of self-healing defects, which appear during construction and building service and design, they are so-called "smart" composites.

3 Results and Discussion

Principles for productivity enhancement of wall materials manufacturing using sandy and clay rocks with an incomplete stage of clay formation and industrial wastes in hydrothermal processing at atmospheric pressure have been suggested. It has been found that aeolian-sedentary-diluvia clay rocks of the Quaternary period are the products of the initial clay formation stage including metastable atelene minerals of nano-level and non-rounded finely dispersed quartz and are suitable as a raw material for autoclaved silicate materials production.

Application of sandy and clay rocks instead of traditional quartz sand in silicate materials production enhances raw mixture formation, hardens raw bricks by 4–11 times, that allows developing the production technique of highly-hollowed construction products.

Application of these rocks allows widening the range of autoclaved raw material base, decrease energy-output ratio of their production, improve the ecological state of the environment and create comfortable conditions for human life.

Within the law of consanguinity it has been found that rocks independent of their genesis (magmatic, methamorphic and aqueous) with banded texture, whose layers are introduced by minerals whose stress related characteristics and thermal-expansion coefficient differ significantly, are non-durable and have anisotropy coefficient 7–9 higher comparing with prototypes with anisotropy coefficient 2–3. Application of the suggested mixtures increases brickwork breaking limit by 3–5 times. It is explained by the contact zone microstructure, for example, ceramic brick and binding matrix. The designed binding matrix and wall material are nearly sole block sample and constructions with a traditional binding matrix have a distinct contact zone – the weakest place of the samples. This law allows creating restoration mixtures, plastering materials, brickwork and restoration compositions of new generations for every walling.

With the account of geonics theoretical background a wide range of acoustic, insulating, and construction insulating composites based on foam glass have been suggested.

Designing construction composites with the account of anthropogenic methasomatosis theory in the construction material science provides possibility of defects self-healing which appear during construction and building service and design so-called "smart" composites.

These materials are designed with the account of the system interacting with the environment. This system allows reacting to the external actions by defects "self-healing" and having positive impact on the triad "a man-material-environment".

This approach has been tested on the composite bindings based on calc-sinter, which create favorable conditions for at the early stages of structure formation and system hardening. It decreases stress in the hardening composite and as a consequence decreases the amount and the size of micro-cracks that predetermines technical and economic efficiency of the composite binding based on tuff, especially in hot climate. Volcanic tuff is known to be a heteroporous rock. Pore space of this rock is rather complex in form and combines pores of different sizes.

Water in this rock is in complex interaction with its mineral grid, whose boundaries and ratio are relative and change constantly: vapour; chemically and physically bounded water; free or gravitational water.

In a hot climate, with a deficit of liquid phase in the concrete itself, tuff particles in the binding mixture composition, during the hardening process will release the capillary water, it will activate structure formation processes and synthesize more massive structure of the materials during concrete hardening and utilization.

The micro-cracks appearing during service at different conditions will self-heal by interaction of water in tuff particles with unreacted cement minerals. During service tuff particles release saved capillary water and that will result in structure formation activation and synthesizing more massive materials structure during concrete hardening and operation. These are so-called smart composites.

4 Conclusions

Hence, monodisciplinary and interdisciplinary approaches in the construction material science promoted developing a wide range of construction composites which main task is construction of solid and durable structures. Design and creation of materials for environment optimization "a man-material-environment" is a complex problem, which requires united work of scientists of different fields. A single way to solve this problem is transdisciplinary approach as a way of widening the scientific world view, which requires considering phenomena beyond a single science.

Acknowledgements. The study is carried out in the framework of the State Task of the RF Ministry of Education and Science No. 7.872.2017/4.6. Development of principles for the design of ecologically positive composite materials with prolonged bioresistance 2017–2019.

References

Ahadi P (2011) Applications of nanomaterials in construction with an approach to energy issue. Adv Mater Res 261–263:509–514

Delgado JMPQ, Cerný R, de Lima AGB, Guimarães AS (2015) Advances in building technologies and construction materials. Adv Mater Sci Eng 2015:1¯3 no. 312613

Elistratkin MY, Kozhukhova MI (2018) Analysis of the factors of increasing the strength of the non-autoclave aerated concrete. Constr Mater Prod 1(1):59–68

Lesovik VS, Chulkova IL, Zagorodnjuk LH, Volodchenko AA, Popov DY (2014) The role of the law of affinity structures in the construction material science by performance of the restoration works. Res J Appl Sci 9(12):1100–1105

Zagorodnyuk L, Lesovik VS, Sumskoy DA (2018) Thermal insulation solutions of the reduced density. Constr Mater Prod 1(1):40–50

32

Alkaline Activation of Rammed Earth Material – "New Generation of Adobes"

C. Costa[1(✉)], D. Arduin[1], C. Sequeira[1], D. Terroso[1], S. Moutinho[2], Â. Cerqueira[1], A. Velosa[1,2], and F. Rocha[1,2]

[1] GeoBioTec, Geosciences Department, University of Aveiro, 3810-193 Aveiro, Portugal
cristianacosta@ua.pt

[2] RISCO, Civil Engineering Department, University of Aveiro, 3810-193 Aveiro, Portugal

Abstract. Adobe is an extremely simple form of earth construction and with this technique the shrinkage associated with the construction of large structures is avoided. In Portugal, earthen materials have been used in load-bearing walls in the form of adobe or rammed earth for the construction of buildings especially in the southern and central coast. Most conventional consolidation treatments used in the past have not succeeded in providing a long-term solution because they did not tackle the main cause of degradation, the expansion and contraction of constituent clay minerals in response to humidity changes. Clay swelling could be reduced significantly by transforming clay minerals into non-expandable binding materials with cementing capacity using alkaline activation. in this study it was being developed adobes with water, and alkaline activated with NaOH and KOH. The obtained results allowed to conclude that the adobes with NaOH and KOH have an increase of its properties.

Keywords: Rammed-earth · Adobe construction · Alkaline activation · "New generation of adobes"

1 Introduction

The use of adobe construction in the Aveiro district reflected the properties of the existing raw materials available to be applied, namely sand, clay sediments and soils and lime (Silveira et al. 2012), and there is an evident heterogeneity of the adobes linked to the geographic distribution of the available resources. In Aveiro there was a semi-industrial production of adobe, some small companies employing 'adobeiros', for the manufacture of blocks of adobe, along with a domestic self-production (Costa et al. 2016). Most conventional consolidation treatments used in the past have not succeeded in providing a long-term solution because they did not tackle the main cause of degradation, the expansion and contraction of constituent clay minerals in response to humidity changes. Clay swelling could be reduced significantly by transforming clay minerals into non-expandable binding materials with cementing capacity using alkaline activation.

2 Methods and Approaches

The chemical composition of diatomite samples was assessed by X-ray fluorescence (XRF), qualitative mineralogical analyses were carried out by X- ray diffraction (XRD). Compressive strength tests use a universal mechanical compression testing machine (Shimadzu Autograph AG 25 TA).

3 Results and Conclusions

Mineralogically, samples are composed by quartz, feldspars and phyllosilicates, however, there is an evidence of the presence of amorphous alumino-silicate phases. Samples with higher mechanical strength are associated with higher specific surface areas, namely, those with NaOH. The alkaline activation promotes the increase of mechanical resistance. With the exception of one sample, all the samples present promising mechanical resistances for their application in rehabilitation works on adobe buildings, since these do not require high resistance, even for the sake of compatibility.

References

Costa CS, Rocha F, Velosa AL (2016) Sustainability in earthen heritage conservation. Sustainable use of traditional geomaterials in construction practice. Geol Soc London, Spec Publ 416. http://doi.org/10.1144/SP416.22

Silveira D, Varum H, Costa A, Martins T, Pereira H, Almeida J (2012) Mechanical properties of adobe bricks in ancient constructions. Constr Build Mater 28:36–44

33

Use of Slags in the Production of Portland Cement Clinker

V. Konovalov(✉), A. Fedorov, and A. Goncharov

Belgorod State Technological University named after V.G. Shukhov, Belgorod, Russia
konovalov52@mail.ru

Abstract. The use of technogenic raw materials as input products in the production of portland cement provides for considerable reduction of energy consumption during clinker burning. The study reveals the mineral formation features caused by the change of the liquid phase composition and crystallization of silicate phases. The use of slags in raw mixes increased the mechanical strength of cements by over 50%.

Keywords: Slag · Cement · Clinker phases · Microstructure

1 Introduction

Any waste can be considered as secondary material resources, which may be fully or partially (as additives) used in production (Klassen et al. 2003). In terms of their physical and chemical properties, slags are similar to igneous rocks used in the production of construction materials. Being exposed to high-temperature treatment and containing basic calcium silicates, they considerably reduce fuel consumption during clinker burning (Vvedensky 1978; Kopeliovich et al. 1998).

2 Methods and Approaches

The study included the methods of chemical analysis of input products and clinkers carried out according to GOST 5382-93. The X-ray phase analysis was performed via the powder diffraction technique using DRON-3 M. The polished sections were studied in reflected light using a universal polarizing microscope NU-2 by Karl Zeiss Jena. The etching of polished sections was carried out via the universal etching agent, i.e. M.I. Strelkov's reagent. The thermal test was carried out using the scanning calorimeter STA 449 F1 Jupiter® by NETZSCH in inert media. Thin slag structures were studied on a scanning electron microscope MIRA3 TESCAN.

3 Results and Discussion

Besides others, the LLC South-Ural Mining and Processing Company utilizes smelter slags as input products. The chemical composition of slags and raw materials given in Table 1 indicates a possibility of replacing the natural constituent, i.e. clay, and slightly reducing the consumption of a carbonate component.

Table 1. Chemical composition of input products, %.

Raw mix components	*CaO*	Al_2O_3	Fe_2O_3	SiO_2	*MgO*	*PPP*
Blast furnace slag	39.70	8.80	2.09	39.08	4.08	0.17
Open-hearth slag	32.36	3.96	21.20	19.76	11.18	3.61
Limestone	52.12	0.24	0.18	1.53	0.48	43.11
Clay	7.57	12.8	6.32	50.45	3.97	12.80

According to X-ray phase analysis, the phase composition of dump blast furnace non-granulated and granulated slag differs only in the intensity of new crystal growths. The main identifiable minerals include gehlenite, akermanit, quartz and melilite minerals.

The DTA method of a dump blast furnace slag in the range of 600–700 °C demonstrates a slight vague exo-effect caused by ferrous iron oxidation. The same temperature range indicates the decomposition of a secondary calcium carbonate followed by the insignificant loss of a sample mass. At 840–900 °C the glass phase is crystallized with further heat release.

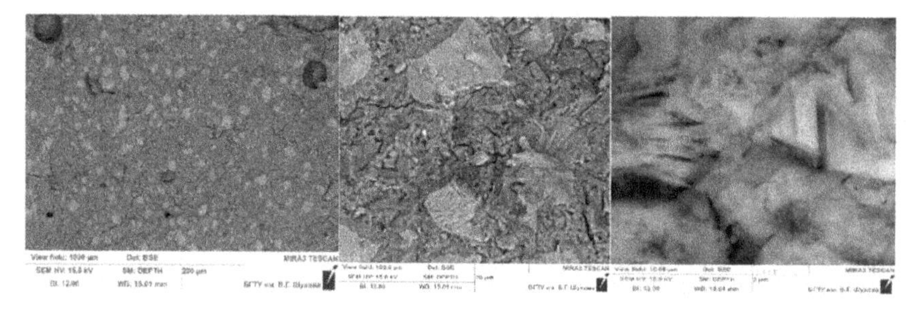

Fig. 1. Microstructure of dump blast furnace slag

The process of devitrification for the blast furnace granulated slag has more expressed exothermic maximum, which is caused by high concentration of a glass phase.

The open-hearth slag acts as a correcting ferrous additive, besides hematite and magnetite is rich in calcium ferrite, monticellite, diopside and magnesium oxide in the form of a periclase.

Figure 1 shows the structure of dump slags representing the conglomerate of crystalline phases and melting particles. The phase formation analysis was carried out in raw mixes having similar chemical composition with various ratio of clay and slag components (Table 2).

Table 2. Raw mixing ratio

Raw mix, No.	*Limestone*	*Clay*	*Blast furnace slag*	*Open-hearth slag*
1	75%	22%	0%	3%
2	61%	0%	33%	6%

The clear exo-effect is observed in a clay-based raw mix (No. 1) at 1227 and 1256 °C caused by the formation of belite phase, as well as endothermal melting effects at 1288 and 1300 °C. According to (Kougiya, Ugolkov 1981), the exotherm of belite mass crystallization at higher temperatures improves the synthesis of alite and its formation in a fine-crystalline state.

Within sintered materials cooled at 1250 °C the crystallization of phase C3A and C4AF is recorded, the reflection intensity of aluminate phase considerably increases with temperature rise. When clay is replaced with slag in a mix the exothermal processes characterizing the formation of silicate phases are weakly expressed within the range of 1000 and 1200 °C. The melting in these mixes is recorded at 1259 and 1308 °C.

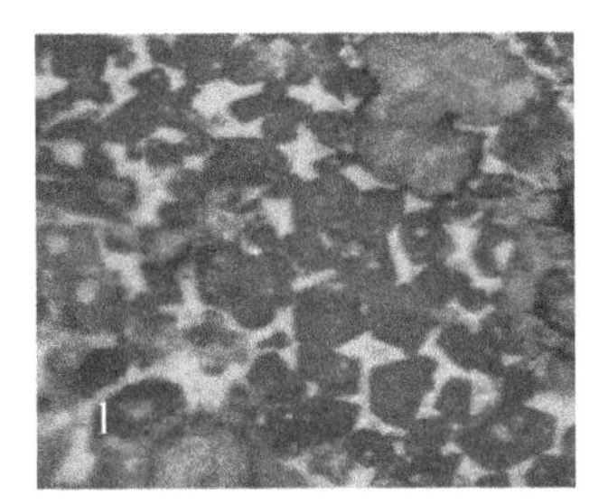

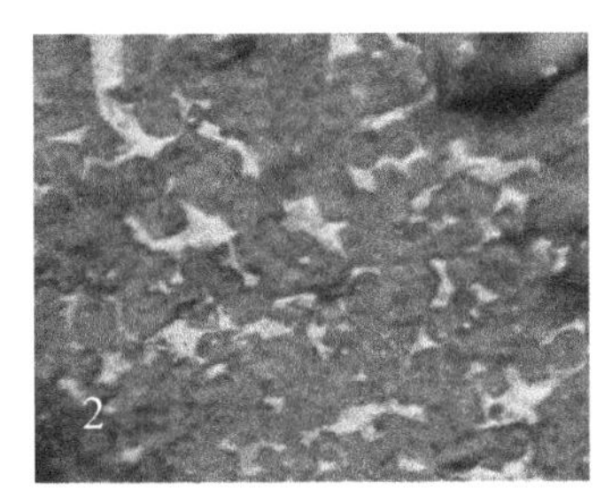

Fig. 2. Micrographs of polished section of sample clinkers: identifications are given in Table 1

Within sintered materials №2 cooled at 1250 °C the aluminate phase prevails thus leading to the appearance of ferrous phases – C6A2F and C4AF with temperature rise. The basic clinker fusion is formed at 1327 °C. These differences in A-F formation are clear with the increase of slag concentration in raw mixes. The gehlenite and mayenite is observed at 1200 °C in clay-containing mixes along with belite phase, which is not observed in slag-containing samples. The introduction of slag intensifies the formation of belite phase at early burning stages. Some changes in the composition of a 'liquid' phase, which increase its temperature and molten viscosity with the increase of slag composition in a raw mix, also affect the features of crystallization of clinker phases.

Figure 2 shows micrographs of polished sections of sample clinkers, which demonstrates the crystallization difference of alite phase.

Clinkers from clay-based raw mixes have clear monadoblastic texture. Clinkers from slag-based raw mixes are different in terms of the number of alite growths.

The strength of cements was defined in small samples from cement paste (1:0) with water-cement ratio of 0.28. The results given in Table 3 show that the use of slags in raw materials instead of clay positively affects the activity of clinkers. The strength improvement within a 28-day interval at full replacement of a clay component with slag made 62%. The heat burning input of limestone-slag mixes may be reduced by over 0.85 mJ/t.

Table 3. Cement stone strength, W/C = 0.28

Mix No., Identifications are given in Table 1	*Tensile strength, MPa, days*		
	2	*7*	*28*
1	7	14	32
2	30	37	52

4 Conclusions

The integral analysis of raw materials and calcined products indicates the possibility of full replacement of clay in a raw mix for the production of portland cement clinker. This contributes to the improvement of qualitative parameters of a calcined product and to the reduction of its production cost.

Acknowledgements. The study is implemented in the framework of the Flagship University Development Program at Belgorod State Technological University named after V.G. Shukhov, using equipment of High Technology Center at BSTU named after V.G. Shukhov.

References

Klassen VK, Borisov IN, Klassen AN, Manuylov VE (2003) Features of mineral formation in slag-containing raw mixes of various basicity. Bull High Educ Inst Constr Ser 7:56–58

Kopeliovich VM, Zdorov AI, Zlatkovsky AB (1998) Utilization of industrial wastes in cement production. Cement 3:174

Kougiya MV, Ugolkov VL (1981) Differential thermal analysis of portland cement raw mixes. Cement 11:19–21

Vvedensky VG (1978) Environmental and economic efficiency of waste utilization. Complex Use Mineral Raw Mater 3:59

34

Genesis of Clay Rock of the Incomplete Stage of Mineral Formation as a Raw Material Base for Autoclaved Materials

A. Volodchenko(✉) and V. Strokova

Belgorod State Technological University named after V.G. Shukhov, Belgorod, Russia
volodchenko@intbel.ru

Abstract. We illustrated possible application of clay rock of incomplete stage of mineral formation to produce autoclaved silica materials. A distinguishing feature of this rock was the presence of thermodynamically unstable compounds. Weathering processes resulted in partial disintegration of the rock, formation of defects in the crystalline lattice of the rock-forming minerals, and partial amorphization of both silicates and aluminosilicates, which reduced the energy potential of the raw materials. Thus, it was possible to usc hydrothermal conditions to boost the formation of neogeneses of cementing compounds and reduce power consumption of the production of autoclaved materials.

Keywords: Lime · Clay rock · Autoclaved silica

1 Introduction

The traditional technology of manufacturing of silica materials is based on using lime and silica sand as raw materials, and the reserves of the latter are getting depleted. Furthermore, this technology is marked with a high power consumption. The consumption of power for grinding the binder and hydrothermal treatment can be reduced by using highly reactive polymineral raw materials, which would allow not only to replace silica sand but also intensify the technological processes. A prospective raw material for producing autoclaved silica materials is the clay rock of the incomplete stage of mineral formation, which is currently unconventional in the building industry (Lesovik 2012; Lesovik et al. 2014; Alfimova and Shapovalov 2013; Volodchenko et al. 2015; Volodchenko et al. 2016; Alfimov et al. 2006; Alfimova et al. 2013). As compared with the traditional silica sand, this rock ensures the synthesis of neogeneses in a more complex system "CaO–[SiO_2–Al_2O_3–(MgO)]–H_2O", which the reduction of power consumption of manufacturing and improving physical, mathematical, and performance properties of both solid and cellular autoclaved materials (Volodchenko and Lesovik 2008a, 2008b; Volodchenko 2011).

The purpose of this paper is to study the genesis of clay rock of the incomplete stage of mineral formation as raw material for producing autoclaved silica materials.

2 Results and Discussion

Some of the most proliferated rocks are deposits of clay. They are formed through weathering of igneous and metamorphic rocks.

The process of weathering of felsic rock passes the following stages: physical weathering (clastic stage), start of the chemical weathering with illite formation (siallite stage), chemical weathering with kaolinite formation (acidic siallite stage), and complete decomposition of silicates with the formation of aluminum, ferrous, and silicon oxides (allite stage). Weathering of rock results in the formation of minerals predominantly of the montmorillonite group.

The processes of weathering can be represented as the destruction of the crystalline structure of the source minerals followed by a transfer through the pseudo-amorphous state and then the formation of the crystalline structure of new rock-forming minerals. However, the ideal amorphous (non-crystalline) state, as well as crystalline state, is not achieved and the transition through it is considered conditional. This results in the change of the source framework structure of the feldspar into the layered structure of clay minerals.

This transition results in rocks that can be characterized as products of the intermediary stage of weathering that occupy a significant stretch in the line of sub-stance transformation and predominate in nature. The rock-forming minerals of these formations include the X-ray amorphous phase, illites, mixed-layer minerals, and imperfectly structured kaolinite and montmorillonite. The formation of these compounds is accompanied by disordering of the crystalline structure of the source minerals, which increases the thermodynamic instability of the rock.

The rocks in the intermediary stage of weathering that were mechanically and chemically activated by exogenous geologic processes, i.e. those that belong to the starting stage of chemical weathering that results in the formation of defects in the structure of silicates and aluminosilicates and the formation of illites, are virtually never used. However, due to their thermodynamic instability, these rocks are the most efficient raw material as components of the systems that must be highly reactive. In particular, such rocks can be used as components of hydro-thermal curing binders, which allows to manage the processes of the formation of the structure of new-generation autoclave materials.

Based on the data about the genesis of the weathering rind rocks, the processes of phase formation in artificial systems of autoclaved curing and how they compare with their natural counterparts of mineral formation, a diagram of the exogenous processes of the genesis of clay rocks as a raw material base for autoclaved materials is proposed (Fig. 1).

The rocks that formed after the second (siallite stage) and partially the third (acidic siallite) stage of weathering belong to the rocks of the incomplete stage of mineral formation that contain compounds of the intermediary stage of weathering: the source minerals have decomposed while the final ones have not formed yet. Such minerals are characterized by the presence of crystalline structure defects and, consequently, thermodynamic instability. This kind of rock can either remain at the place of its formation

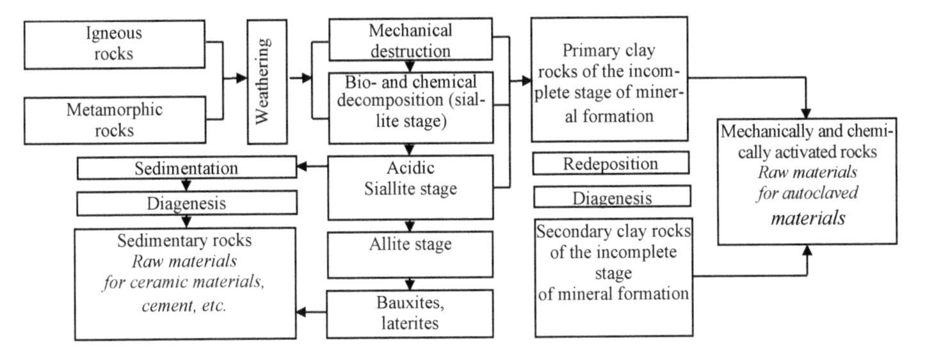

Fig. 1. Diagram of the exogenous processes of the genesis of clay rocks as raw materials base for autoclaved materials

or redeposit in new places as the result of transfer and diagenesis. These rocks are useful as raw materials for autoclaved curing materials.

Thus, weathering of source rocks of different compositions leads to the formation of mineral resources that are different in their composition and industrial significance. If one considers the weathering rind profile, it can be seen that at the moment the building materials industry is using the lower horizon (source rocks) and products of the final stage of weathering: kaolinite and montmorillonite clays as well as bauxites.

Clay rocks of the incomplete stage of mineral formation are some of the most inconsistent in terms of composition and the structure of the sedimentation mass deposits. In terms of the crystallochemical features of the rock-forming minerals, it is necessary to note the high content of the crystalline structure defects, the dis-order of the interlayer aluminosilicate stacks in the layered structure of minerals, etc. The crystalline lattice of the typical clay minerals that belong to layered silicates is characterized by the alteration of silicon-oxygen tetrahedra and hydroxyl octahedra. To some degree, clay minerals can be considered objects, where the nanoparticles (elementary layer packets) are packed into the microscopic structures, where the properties of the individual constituent particles are concealed due to a strong interparticle interaction.

3 Conclusions

Thus, lay rocks of the incomplete stage of mineral formation are energy-saturated deposits. Due to the composition and crystallochemical features of the rock-forming minerals, these rocks can be recommended as energy-efficient raw materials for the production of autoclaved curing materials. A distinguishing feature of this rock is the presence of thermodynamically unstable compounds. At the same time, aluminosilicates are characterized by variable chemical composition and the imperfection of their crystalline lattice. Weathering processes result in partial disintegration of rock, formation of defects in the crystalline lattice of the rock-forming minerals, and partial amorphization of both silicates and alumino-silicates, which reduces the energy potential of the rock-forming minerals. Thus, it is possible to use hydrothermal conditions to boost the formation of neogeneses of cementing compounds and reduce the power consumption of the production of autoclaved materials.

Acknowledgements. The work is realized in the framework of the Program of flagship university development on the base of the Belgorod State Technological University named after V.G. Shukhov, using equipment of High Technology Center at BSTU named after V.G. Shukhov.

References

Alfimov SI, Zhukov RV, Volodchenko AN, Yurchuk DV (2006) Technogenic raw materials for silicate hydration hardening. Mod High Technol 2:59–60

Alfimova NI, Shapovalov NN (2013) Materials autoclaved using man-made aluminosilicate materials. Fundam Res 6(3):525–529

Alfimova NI, Shapovalov NN, Abrosimova OS (2013) Operational characteristics of silica brick, manufactured using man-made aluminosilicate materials. Bulletin of BSTU named after V.G. Shukhov, vol 3, pp 11–14

Lesovik VS (2012) Geonik. Subject and tasks. Publisher Belgorod State Technological University. VG Shukhov, p 219

Lesovik VS, Volodchenko AA, Svinarev AA, Kalashnikov NV, Rjapuhin NV (2014) Reducing. World Appl Sci J 31(9):1601–1606

Volodchenko AA, Lesovik VS, Volodchenko AN, Glagolev ES, Zagorodnjuk LH, Pukharenko YV (2016) Int J Pharm Technol 8(3):18856–18867

Volodchenko AA, Lesovik VS, Volodchenko AN, Zagorodnjuk LH (2015) Int J Appl Eng Res 10(24):45142–421149

Volodchenko AN, Lesovik VS (2008a) Increasing production efficiency of autoclave materials. In: Proceedings of the higher educational institutions. Building, vol 9, pp 10–16

Volodchenko AN, Lesovik VS (2008b) Autoclave silicate materials with nanometer-sized materials. Build Mater 11:42–44

Volodchenko AN (2011) Features magnesia clay interaction with the calcium hydroxide in the synthesis and the formation of tumors microstructure. Bulletin of Belgorod State Technological University. VG Shukhov, vol 2, pp 51–55

35

Estimation of Rheo-Technological Effectiveness of Polycarboxylate Superplasticizer in Filled Cement Systems in the Development of Self-Compacting Concrete for High-Density Reinforced Building Constructions

T. Nizina(✉), A. Balykov, V. Volodin, and D. Korovkin

Department of Building Structures, Ogarev Mordovia State University, Saransk, Russia
nizinata@yandex.ru

Abstract. Analysis of rheo-technological effectiveness of the polycarboxylate superplasticizer Melflux 5581 F in the microcalcite-filled cement systems was performed. Optimal quantities of polycarboxylate superplasticizer and carbonate filler were determined, which allow obtaining highly mobile cement-mineral suspensions at reduced water content, which is an important step in the development of self-compacting concrete mixtures.

Keywords: Cement-mineral suspension · Rheology · Efficiency · Superplasticizer · Microcalcite · Self-compacting concrete

1 Introduction

Currently, there is an active growth in the area of concrete development, which the world technological community clearly classifies as cement composites of the new generation with high strength, workability, volume stability and durability (Collepardi 2006; Nawy 2001; Nizina and Balykov 2016; Nizina et al. 2017; Sivakumar et al. 2014; Tran and Kim 2017). A special place among the concretes of the new generation is occupied by Self-Compacting Concrete (SCC) – Selbstverdichtender Concrete (SVB, German), Betonautoplacant (BAP, French), which are currently fairly widespread abroad. This term, proposed in 1986 by the Japanese professor H. Okamura (Okamura and Ouchi 2003), combines concrete mixtures with high workability characteristics (standard cone flow over 55–60 cm at water-to-cement ratio reduced to 0.35–0.4 or less), which are due to the high deformability of the suspension matrix, along with its high resistance to segregation or separation during movement.

A number of papers experimentally proved that one of the basic principles for producing self-compacting concrete mixtures is the presence of a significant amount of dispersed micro-particles of cement or mineral fillers (mainly 1–100 microns in size) in their formulation, which, together with Portland cement, increase the volume of a dispersion-water suspension. At the same time, not all dispersed fillers are capable of providing a higher flowability in suspension with a superplasticizer (SP) as compared with cement suspensions. The research (Kalashnikov et al. 2014) shows that carbonate rocks (limestone, marble, dolomite), which include particles with a significant proportion of positively charged active centers, are the most compatible with anionic superplasticizers.

Thus, the generate of self-compacting concrete mixtures must begin with the development of rheological active formulation of the filled cement binders, which allow, when mixing together with the superplasticizer, to form aggregate resistant suspensions, which have a high concentration of the solid phase, low values of the shear stress limit and plastic viscosity at high gravitational fluidity under its own weight. At the same time, the efficiency of plasticizers in such systems will depend on many factors – addition procedure and optimal quantity of the plasticizer, rheological activity of the fillers used, etc.

The purpose of the study is to determine the optimal quantity of polycarboxylate superplasticizer and carbonate filler in cement-mineral suspensions when developing self-compacting concrete mixtures.

2 Methods and Approaches

To prepare suspensions, Portland cement CEM I 32.5R (C) by Mordovtsement PJSC (GOST 31108) was used. The mineral part included a carbonate filler from KM100 microcalcite (MKM) by Polipark LLC with the dosage of 0÷300% of Portland cement weight (0÷75% of solid phase weight) with the variability pitch of 100%. Melflux 5581 F (SP) polycarboxylate superplasticizer by BASF Construction Solutions (Trostberg, Germany) was used as a plasticizer.

The study was carried out with the fixed water-solid ratio W/S = 0.15 with the varying factors being:

- ratio MKM/C, x_1 = 0÷3.0 relative units;
- ratio SP/(C + MKM), x_2 = 0÷1.5%.

3 Results and Discussion

The study results were used to develop an experimentally statistical (ES) model describing the changes in the flow diameter of cement-carbonate suspensions (Portland cement + microcalcite) [D_f^{HC}, mm] from the Hegermann cone from the content of the varying factors x_1 and x_2:

$$\begin{aligned} D_f^{HC} &= 288.9 + 52.9 \cdot x_1 + 64.8 \cdot x_2 + 19.8 \cdot x_1 \cdot x_2 - 51.33 \cdot x_1^2 \\ &\quad -89.3 \cdot x_2^2 + 37.97 \cdot x_1^3 + 32.34 \cdot x_2^3 - 59.06 \cdot x_1 \cdot x_2^2 \\ &\quad -15.19 \cdot x_1^2 \cdot x_2 + 31.64 \cdot x_1^2 \cdot x_2^2. \end{aligned} \tag{1}$$

Using the polynomial (1), isolines have been built, which reflect the changes in the flow diameter of cement-carbonate suspensions from the Hegermann cone depending on the content of microcalcite and Melflux 5581 F superplasticizer (Fig. 1). It has been found that for the constant water/solid ratio of W/S = 0.15 relative units, an increase in the dosage of the superplasticizer and the mineral filler (microcalcite) causes a significant increase in the flow diameter of the cement-mineral suspension.

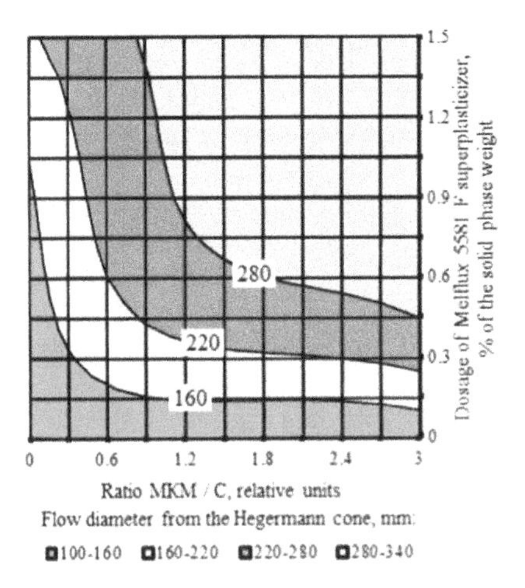

Fig. 1. Changes in the flow diameter of cement-carbonate suspensions from the Hegermann cone depending on the superplasticizer dosage and microcalcite filling degree

Analysis on Fig. 1 shows that cement suspensions without microcalciteat the specified water content W/S = 0.15 begin to flow under the action of gravity and the value of the Hegermann cone flow diameter $D_f^{HC} = 100 \div 135$ mm with superplasticizer quantities exceeding 0.1÷0.5% the mass of the solid phase. However, even an increase in the superplasticizer content to 1.0÷1.5% does not allow achieving self-compacting of suspensions ($D_f^{HC} = 160 \div 210 < 280$ mm).

It was determined (Fig. 1) that the level of rheo-technological indices specified for self-compacting suspensions ($D_f^{HC} \geq 280$ mm) is achieved at the microcalcite content $x_1 = 0.82 \div 3.0$ relative units (82÷300% of the Portland cement mass or 45÷75% of the solid phase mass (C + MKM)) and the quantity of Melflux 5581 F superplasticizer $x_2 = 0.45 \div 1.5\%$ of the solid phase mass, wherein when decreasing the indicator x_1, an increase in the indicator x_2 in the specified ranges (($x_1 = 3.0$ relative units; $x_2 = 0.45\%$) → ($x_1 = 0.82$ relative units; $x_2 = 1.5\%$)) is required.

4 Conclusions

The analysis of research results found the optimal levels of the varying factors that allow reaching the self-compacting of suspensions for the flow diameter from the Hegermann cone above 280 mm and the water/solid ratio of 0.15 relative units: the dosage of Melflux 5581 F superplasticizer is 0.5÷1.0% of the solid phase weight; the filling degree of suspension with microcalcite is at least 105% of the Portland cement weight.

Acknowledgements. The reported study was funded by RFBR according to the research project № 18-29-12036.

References

Collepardi M (2006) The New Concrete. Grafiche Tintoretto, Villorba

Kalashnikov VI, Moskvin RN, Belyakova EA, Belyakova VS, Petukhov AV (2014) High-dispersity fillers for powder-activated concretes of new generation. Syst Methods Technol 2 (22):113–118

Nawy EG (2001) Fundamentals of High-Performance Concrete. Wiley, New York

Nizina TA, Balykov AS (2016) Experimental-statistical models of properties of modified fiber-reinforced fine-grained concretes. Mag Civil Eng 2:13–25. https://doi.org/10.5862/MCE.62.2

Nizina TA, Ponomarev AN, Balykov AS, Pankin NA (2017) Fine-grained fibre concretes modified by complexed nanoadditives. Int J Nanotechnol 14:665–679. https://doi.org/10.1504/IJNT.2017.083441

Okamura H, Ouchi M (2003) Self-compacting concrete. J Adv Concr Technol 1:5–15

Sivakumar N, Muthukumar S, Sivakumar V, Gowtham D, Muthuraj D (2014) Experimental studies on high strength concrete by using recycled coarse aggregate. Res Inventy: Int J Eng Sci 4:27–36

Tran NT, Kim DJ (2017) Synergistic response of blending fibers in ultra-high-performance concrete under high rate tensile loads. Cement Concr Compos 78:132–145

36

Influence of Flow Blowing Agent on the Properties of Aerated Concrete Variable Density and Strength

V. Galdina(✉), E. Gurova, P. Deryabin, M. Rashchupkina, and I. Chulkova

Department of Building Structures, Ogarev Mordovia State University, Saransk, Russia
example@yandex.ru

Abstract. We presented the effect of gasifier flow on gas release kinetics and basic properties of aerated concrete with variable traverse density and strength. We found optimal consumption of gasifier for manufacturing of aerated concrete with variatropic properties from expanded clay sand, wherein we obtained aerated concrete with a strength of 1.42 MPa at an average density of 414 kg/m^3.

Keywords: Cellular concrete · Aerated concrete · Blowing consumption · Bloating · Gas release kinetics

1 Introduction

The scientists and technologists should develop a technology of new generation of cellular concrete with a higher strength and frost-resistance at a low average density.

2 Methods and Approaches

The production of concrete products in a closed mold can be related to new methods. Works by Chernov and Zavadsky are important at the present stage of production of aerated concrete. Their studies are mainly based on preparation of gas concrete mix in the mold with a hollow cap (without holes), or small holes in side and top faces of the mold (Chernov et al. 1983; Chernov 2003; Zavadsky 2005; Zavadsky 2001; Chernov 2002; Zavadsky, Kosach 1999).

The properties of aerated concrete with variable traverse density and strength of products are influenced by the following factors: area of surface of a cover; fluidity of mix; consumption and type of blowing agent; fill level of mix in the mold; type, flow rate and surface area of silica component and binder.

3 Results and Discussion

The formation of a cellular structure and gas-concrete products in an individual mold was carried out as follows: gas-concrete mix was prepared, poured into a mold closed by a cover with a circular hole in the center of the mold (The method…, 2015).

The essence of formation of a cellular structure of concrete of variable traverse density and durability is as follows. Hydrogen releases when an alkaline component binds with aluminum, which results in blowing of viscoplastic mass. The blowing mix, having reached the internal surface of the cover with a circular hole, meets a barrier on the way and swells up on the way of the weakest resistance (through the hole), and on the periphery of the product there is a self-consolidation of aerated concrete due to overpressure. As a result, the pressing gas with more than 0.1 kgf/cm^2 and presence of the closed cover surface at the time of a blowing results in less dense and more porous products in the mid-range of the composite and denser, stronger – on the periphery.

The known formulations in production of gas-concrete products by traditional technology, in particular a gas developing agent consumption, are not absolutely correct for manufacturing of products in the closed mold since the raised gas developing agent consumption is necessary for effective self-consolidation of exemplars. In this regard it is necessary to reveal its influence on a kinetics of gas emission of the mix and main properties of aerated concrete.

V/T = 0.5 was applied to aerated concrete on the basis of expanded clay sand with mix spread at Southard viscosimeter 30 cm. Gazobetolyuks gas paste and earlier chosen area of the closed cover surface with the round section equal to 71% was applied as gas developing agent. The mix was poured up to 70% of the mold height.

The nature of a flatulence of the mix up to the cover in the period of time from 5 to 20 min at all gas-concrete mix was identical, different only by the amount of gas developing agent. Process of an aerogenesis in all mixtures practically finishes in 20–30 min (Fig. 1).

The highest limit of compression strength is observed for samples produced at a gasifier flow 1000 g per m^3 of the mix. As a result of the chemical reaction the released hydrogen swells the mix which reaches the edge of the mold, where it meets an obstacle of cover and is pressed on the way of the weakest resistance through a circular hole in the cover, whereby mostly form crusts and increase strength of samples during compression. When the flow reduces to 800 g the strength is reduced on average by 30%. This is a result of low flow of the gasifier, and the mix during swelling is not sufficient for self-sealing along the periphery, which is confirmed by gas release kinetics, which reaches the inner surface of the cover only in 20–22 min and the final height of swelling 35%. Since one of the main objectives is not only to obtain aerated concrete with a relatively high compressive strength, but also with reduced average density index, the optimum flow rate is from 1100 to 1300 g per m^3.

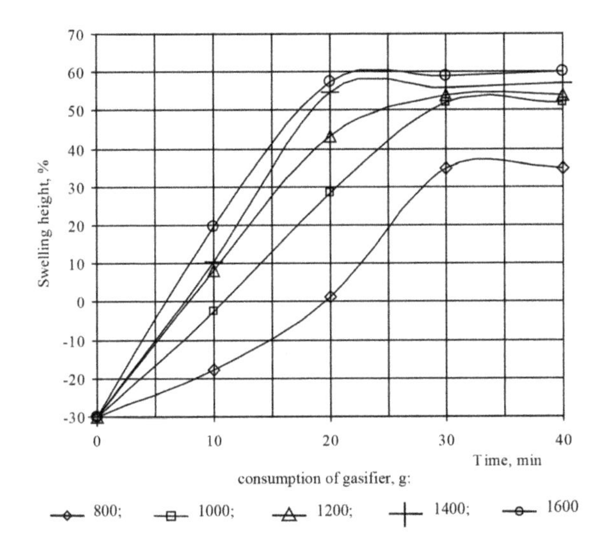

Fig. 1. Kinetics of gas emission of mix on the basis of expanded sand

Aerated concrete, prepared in the mold with a circular hole in the cover on expanded clay aggregate with the strength 1.42 MPa at an average density 414 kg/m^3 (Fig. 2), was obtained at a flow rate equal to 1,200 g of blowing agent per m3 of mix, which corresponds to D500 brand strength. The gassing process also takes about 25–30 min, after which the height of the crusts is 50–54% (Fig. 2).

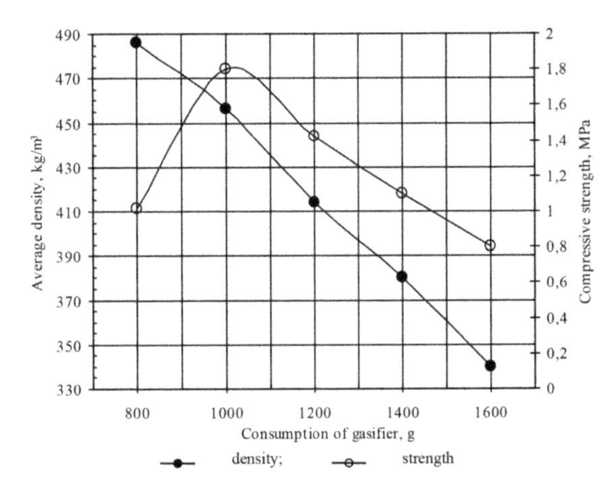

Fig. 2. The effect of the gasifier consumption on the average density and strength of aerated concrete based on expanded clay sand

4 Conclusions

We determined the optimal flow of gas agent on the basis of expanded clay sand equal to 1100–1300 g per m^3, at which the aerated concrete with a durability of 1.42 MPa at an average density of 414 kg/m^3 that corresponds to D500 brand was produced. At the same time the process of aerogenesis continues 25–30 min, after which the height of top crust is 50–54%.

Advantage of technology of aerated concrete with variatropic properties is a high stability of density of cellular concrete depending only on the accuracy of the gravimetric feeders measuring components. At the same time not only density, but also durability, deformability, heat conductivity and other properties are stabilized. The increased dispersion of operational parameters, inherent to all cellular concretes, is excluded.

Development of new and combination of already known methods of pore making in mass and also application of processing methods for products with variatropic properties at formation allow producing one and two-layer wall products of various configuration, sizes and required heat-shielding and operational properties.

References

Chernov AN (2003) Auto-frettage in aerated concrete technology. Constr Mater 11:22–23

Chernov AN (2002) Cellular concretes. Publishing house of SUSU, Chelyabinsk, 111 p

Chernov AN, Kozhevnikova LP, Khmelev SV, Tsarkov VV, Danilyuk MA, Moiseev EI, Stepanova ZA (1983) Technology of cellular concrete products with compacted surface layer. Constr Mater 8:12–13

The method of producing aerated concrete and the raw mix for its preparation. Patent of the Russian Federation P. Deryabin. No. 2560009 of July 20 2015

Zavadsky VF (1999) An integrated approach to solving the problem of thermal protection of walls of heated buildings. Constr Mater 2:7–8

Zavadsky VF (2001) Variants of wall constructions with the use of effective heaters - Novosibirsk: NGASU, 52 p

Zavadsky VF, Kosach AF (2005) Wall materials and products. SibADI Publishing House, Omsk, 254 p

37

Properties Improvement of Metakaolin-Zeolite-Diatomite-Red Mud Based Geopolymers

F. Rocha[1(✉)], C. Costa[1], W. Hajjaji[1], S. Andrejkovičová[1], S. Moutinho[2], and A. Cerqueira[1]

[1] GeoBioTec, Geosciences Department, University of Aveiro, 3810-193 Aveiro, Portugal
tavares.rocha@ua.pt

[2] RISCO, Civil Engineering Department, University of Aveiro, 3810-193 Aveiro, Portugal

Abstract. Addition of pozzolanic materials increases the mechanical characteristics of construction materials and contributes towards a higher durability. Metakaolin is an artifical pozzolan obtained by calcination of kaolinitic clays at an adequate temperature. Geopolymers are inorganic materials from mineral origin, composed of a precursor, an alkaline activator and a solvent. New geopolymer formulations were designed by sodium silicate/NaOH/KOH activation of metakaolin, zeolites, diatomites and red mud mixtures. The effects of source materials on the microstructure and mechanical properties were studied. Mineralogical and chemical compositions were assessed as well as microstructure, specific surface area, compressive strength and adsorption. In general, incorporation of red mud, zeolite filler and diatomites to metakaolin in medium of alkali activators of low concentration provided formation of more eco-friendly materials with high mechanical resistances and water treatment capabilities.

Keywords: Geopolymers · Metakaolin · Alkaline activation · Additives formulations

1 Introduction

Addition of pozzolanic materials increases the mechanical characteristics of construction materials and contributes towards a higher durability. Metakaolin is an artifical pozzolan obtained by calcination of kaolinitic clays at an adequate temperature. Metakaolin is the structurally disordered product obtained following the dehydroxylation of kaolin, more precisely of its essential component kaolinite - $Al_2Si_2O_5(OH)_4$.

Geopolymers are inorganic materials from mineral origin, composed of a precursor, an alkaline activator and a solvent. The process of geopolymerization involves a chemical reaction that takes place in an alkaline medium, resulting in the formation of inorganic polymers that have silicon and aluminum as main constituents (Si+Al), connected by oxygen ions. The solution becomes alkaline using activators such as sodium and potassium hydroxides and/or sodium and potassium silicates.

Depending on the composition, geopolymer characteristics can be attained in terms of physical, chemical and mechanical performance (Mackenzie and Welter 2014). Moreover, geopolymers have the advantage to be possibly formulated from a wide range of aluminosilicate minerals.

2 Methods and Approaches

New geopolymer formulations were designed by sodium silicate/NaOH/KOH activation of metakaolin, zeolites, diatomites and red mud mixtures. The effects of source materials on the microstructure and mechanical properties were studied.

Geopolymers were prepared using commercial metakaolin (1200S, AGS Mineraux, France, D50 = 1.1 μm, bulk density = 296 g/dm^{-3}), diatomite from Rio Maior and Amieira deposits (Portugal), zeolite (ZeoBau micro 50, from Nižný Hrabovec, Zeocem, Slovakia, CEC = 83 meq/100 g, SSA = 1663 m^2/kg, particle size 0–0.05 mm, bulk density = 500–600 g/dm^{-3}, more information about Nižný Hrabovec deposit is available on http://www.iza-online.org/natural/), red mud from aluminium metallurgies wastes, hydrated sodium silicate (Merck, Germany Merck, Germany; 8.5 wt% Na_2O, 28.5 wt% SiO_2, 63 wt% H_2O) and sodium and potassium hydroxide (ACS AR Analytical Reagent Grade Pellets).

The role of the above ingredients in the preparation of geopolymers is as follows: metakaolin was used as a precursor of aluminium; diatomite was used as a precursor of silica; zeolite was used as filler with high specific surface area and cation exchange capacity; red mud as precursor of aluminium and iron; sodium silicate was used as a source of silicon and sodium and potassium hydroxide as alkaline activators for dissolution of aluminosilicate. Water was the reaction medium.

The X-ray diffraction was conducted on Philips X'Pert diffractometer using CuKα radiation at a speed of 0.02°/s. The X'Pert HighScore (PW3209) program was used to analyze XRD peaks. The chemical composition (major elements) of materials was analysed using PANalytical Axios X-ray fluorescence spectrometer. Loss on ignition was determined by heating the samples in an electrical furnace at 1000 °C during 3 h. The microstructural characterization was carried out by scanning electron microscopy (SEM – Hitachi, SU 70) and energy dispersive X-ray spectrometry (EDS – EDAX with detector Bruker AXS, software: Quantax) operated at 3–30 kV.

The specific surface area of geopolymers (heated at 200 °C during 6 h) was determined by BET method (Brunauer, Emmett and Teller, nitrogen gas adsorption at 77 K) (Brunauer et al. 1938). Compressive strength tests were carried out on 3 probes of individual geopolymer on (SHIMADZU: AG-IC 100 kN) equipment, by applying a maximum force of 5 kN at the speed of 50 N/s according to the Standard EN 1015-11. Adsorption experiments were carried out in batch using nitrate solutions of Pb^{2+}, Zn^{2+}, Cu^{2+}, Cd^{2+} and Cr^{3+}.

3 Results and Conclusions

Performed analyses confirm that zeolite particles are responsible for higher amount of crystalline phases producing more compact and firm microstructure of blended geopolymers relative to that of reference MK100. SEM analysis reveals that incorporation of 50 wt% of zeolite to metakaolin geopolymer (MK50) leads to the denser geopolymer matrix related to microstructure when compared to 100 wt% metakaolin in reference geopolymer (MK100). Dense microstructure of MK50 manifests in its considerably lower water adsorption and specific surface area values.

All geopolymers containing zeolite show increase in compressive strength compared to pure metakaolin one, with optimal ratio metakaolin precursor/zeolite filler 50:50 providing the highest compressive strength (8.8 MPa at 28 days). The adsorption of heavy metals increases as the amount of metakaolin in the structure increases.

Diatomite enriched formulations showed different evolutions according to the Na or K activator; those K activated showed higher compressive strength.

The well-known geopolymer composition (amorphous phase predominant over residual quartz, illite and anatase) is slightly affected by the red mud, despite it provides aluminum and iron oxides and oxy-hydroxide (hematite, goethite, gibbsite and boehmite). These phases, in fact, are to a limited extent involved in the geopolymerization process and the alkaline aluminosilicate amorphous phase is about 90 wt% in all samples. The structural features of the amorphous geopolymer, as resumed by the broad hump of XRD patterns, are not modified by addition of RM.

The physical properties of geopolymers are not significantly affected by RM, as all samples exhibit high values of water absorption and low apparent density. The chemical stability is good: sodium leaching test gave leachate concentrations close to 100 ppm without evidences of deterioration of mechanical performance. Red mud influences the mechanical strength during curing (especially at the higher amounts of RM but the lower additions, cured for 28 days showed good compressive strength).

In general, incorporation of red mud, zeolite filler and diatomites tometakaolin inmedium of alkali activators of low concentration provided formation of more eco-friendly materials with high mechanical resistances and water treatment capabilities.

References

Brunauer S, Emmett PH, Teller EJ (1938) Adsorption of gases on multimolecular layers. J Am Chem Soc 60:309–319

Mackenzie K, Welter M (2014) Geopolymer (aluminosilicate) composites: synthesis, properties and applications. Adv Ceram Matrix Compos. 445–470

38

Influence of Sodium Oxide on Brightness Coefficient of Portland Cement Clinker

D. Mishin(✉) and S. Kovalyov

Department of Technology of Cement and Composite Materials,
Chemical Technology Institute, Belgorod State Technological University
named after V.G. Shukhov, Belgorod, Russia
mishinda.xtsm@yandex.ru

Abstract. The paper is devoted to the possibility of adjusting the reflection factor of portland cement clinker. For this purpose, Na_2CO_3 is introduced into the slurry of CJSC Belgorod Cement Plant. The influence of Na_2O on brightness coefficient of the crushed clinker is established at a burning temperature of 1250–1300 °C. With the increase of Na_2O concentration up to 1–2% the brightness coefficient is reduced and the increase of Na_2O in the range of 3.5–7% leads to sharp increase of the reflection factor and CaO_{free} content.

Keywords: Sodium oxide · Brightness coefficient · White cement · Free calcium oxide

1 Introduction

Modern rotary furnaces of cement industry are characterized by accumulation and circulation of alkali salts in the furnace system (Luginina 2002). As a result, the R_2O content of furnace charge may reach 3.5% before the sintering zone, and in some plants this value may even reach 10%. The formation of calcium aluminate ferrite is not observed in furnace charges of alumina industry characterized by high Na_2O content due to the formation of aluminates and sodium ferrites (Lisiyenko 2004). Hence, as may be expected, the accumulation and circulation of alkali salts will lead to the situation that occurs in furnaces of alumina industry.

Raw mixes from analytical reagents used to obtain C_4AF are characterized by the formation of sodium compounds instead of calcium compounds (Kovalyov 2015). It will allow obtaining a clinker of lighter shade.

Thus, the purpose of this study is to determine the possibility of adjusting the brightness coefficient of portland cement by introducing sodium oxide into the raw mix.

2 Methods and Approaches

The dried slurry of CJSC Belgorod Cement Plant with the following modular characteristics: KH = 0.91; n = 2.23; p = 1.29 was used as a raw mix (Table 1).

Table 1. Chemical composition of slurry of CJSC Belgorod Cement Plant, %

Losses on ignition	*SiO_2*	*Al_2O_3*	*Fe_2O_3*	*CaO*	*MgO*	*K_2O*	*Na_2O*	*SO_3*	*Other*
34.8	14.23	3.59	2.78	43.12	0.6	0.4	0.11	0.09	0.37

Sodium carbonate of CH classification was introduced into the raw mix in the amount of 0.5; 1; 2; 3.5; 5; 7% of Na_2O in equivalent of ignited basis. Tablets weighing 2 g were formed from obtained mix under manual pressure. In order to avoid Na_2O volatilization during roasting, the samples were covered with a platinum cup. Roasting of samples was carried out in a laboratory furnace with isothermal time of 20 min. The heating rate of the furnace made 10 °C/min.

The influence of alkali salts on the formation of aluminate-ferrite phase was estimated according to the content of free calcium oxide and the brightness coefficient of samples. The CaO_{free} content in a clinker was defined through ethyl-glycerate method (Butt and Timashev 1973). The whiteness (brightness coefficient) of a clinker was defined via FB-2 reflection meter on a reference polished barium sulfate plate.

3 Results and Discussion

The compositions burned at 1250 and 1300 °C were analyzed. The roasting temperature was chosen based on the following: at given temperatures the formation of belite was complete, and the synthesis of alite had not started yet. As a result, the formation of silicate phases would not change the content of free calcium oxide.

At a temperature of 1250 °C (Fig. 1) at the initial stage (up to 1% Na_2O) the content of free calcium oxide is reduced alongside with the brightness coefficient (BC). The brightness coefficient almost does not change with Na_2O concentration in the range of 1–3.5% and the content of free calcium oxide. With the introduction of over 3.5% of Na_2O the CaO_{free} content and the brightness coefficient sharply increase.

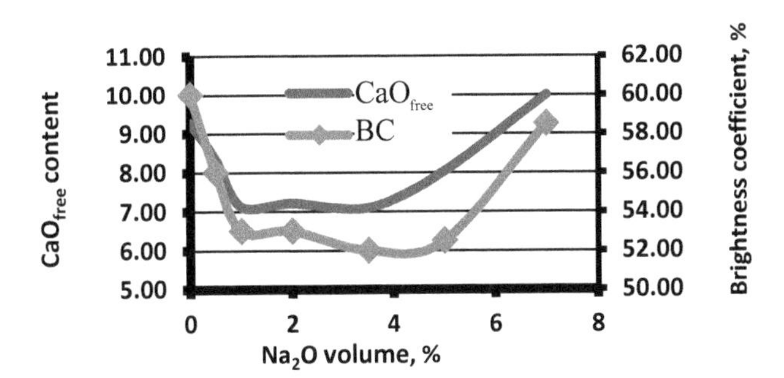

Fig. 1. Influences of Na_2O on brightness coefficient and content of free calcium oxide at 1250 °C roasting temperature.

At a roasting temperature of 1300 °C (Fig. 2) a similar situation is observed. With the amount of introduced Na_2O up to 1–2% the considered characteristics are reduced. Sharp increase of reflection factor and CaO_{free} content was also observed in the range of 3.5–7% of Na_2O.

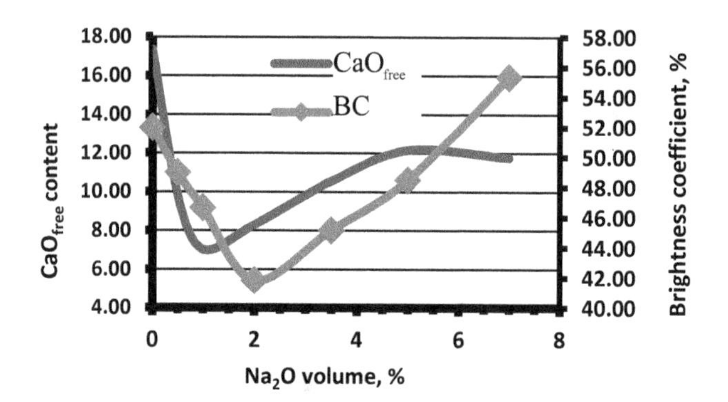

Fig. 2. Influence of Na_2O on reflection factor and content of free calcium oxide at 1300 °C roasting temperature.

The relatively low content of free calcium oxide without alite formation may be caused by the formation of intermediate phases, for example carbonate spurrite. The presence of Na_2CO_3 in the system fosters its formation. Hence, the increase of Na_2O content at the initial stage leads to CaO_{free} decrease.

The sharp increase of free calcium oxide content starting from Na_2O concentration of approximately 3.5% confirms the formation of ferrites or sodium aluminates instead of calcium aluminate ferrites in clinker systems, since such interaction results in the emission of additional amounts of calcium oxide and may be presented by the following reactions:

$$3CaO \cdot Al_2O_3 + Na_2O = 2NaAlO_2 + 3CaO \tag{1}$$

$$4CaO \cdot Al_2O_3 \cdot Fe_2O_3 + 2Na_2O = 2NaFeO_2 + 2NaAlO_2 + 4CaO \tag{2}$$

The change dependence of brightness coefficient within samples on Na_2O content in a clinker correlates with the change dependence of free calcium oxide content. This may be explained by the fact that CaO_{free} crystals are white. Therefore, the increase or reduction of its content leads to the corresponding change of the brightness coefficient of a clinker. Based on above conclusions, the sharp increase of the brightness coefficient with the increase in Na_2O concentration of over 3.5% may also be caused by the complexity of aluminate-ferrite phase. This phase is the most colored part of a clinker (Zubekhin et al. 2004). Hence, the change of its composition and quantity resulting from the formation of aluminates and sodium ferrites will be characterized by the corresponding change of the brightness coefficient of a clinker.

4 Conclusions

The introduction of up to 2–3.5% of Na_2O leads to the decrease in the brightness coefficient of samples due to the reduction of free calcium oxide content. Apparently, this is caused by the formation of intermediate phases – carbonate spurrite.

The Na_2O content of over 3.5% increases the brightness coefficient of samples. This is caused by the formation of aluminates and sodium ferrites instead of aluminates and calcium aluminate ferrites.

Acknowledgements. The study is implemented in the framework of the Flagship University Development Program at Belgorod State Technological University named after V.G. Shukhov, using the equipment of High Technology Center at BSTU named after V.G. Shukhov.

References

Butt YM, Timashev VV (1973) Practical Guide on Chemical Technology of Binding Materials: Study Manual. Higher School, p 504

Zubekhin AP, Golovanova SP, Kirsanov PV (2004) White portland cement. Publishing House of Izvestiya Vuzov Journal, Rostov-On-Don, North Caucasian Region, p 264 (Zubekhin AP (ed))

Kovalyov SV (2015) Essentially new method of clinker bleaching with high iron content. In: Kovalyov SV, Mishin DA (eds) International seminar-competition of young scientists and graduate students working in the field of binding materials, concrete and dry mixes: collection of articles: 180 pages. Alitinform Publishing House, Saint Petersburg, pp 29–37

Lisiyenko VG (ed) (2004) Rotary furnaces: thermal engineering, management and ecology. Teplotekhnik, p 688 (Reference edition. Lisiyenko VG, Shchelokov YM, Ladygichev MG)

Luginina IG (2002) Selected articles. Publishing House of BSTU named after V.G. Shukhov, Belgorod, p 302

Biomimetic Materials on a Mineral Basis, Biomineralogy

Biomimetic Superhydrophobic Cobalt Blue/Clay Mineral Hybrid Pigments with Self-Cleaning Property and Different Colors

B. Mu[1], A. Zhang[1,2], and A. Wang[1(✉)]

[1] Key Laboratory of Clay Mineral Applied Research of Gansu Province, Center of Eco-material and Green Chemistry, Lanzhou Institute of Chemical Physics, Chinese Academy of Sciences, Lanzhou 730000, People's Republic of China
aqwang@licp.cas.cn

[2] Center of Materials Science and Optoelectronics Engineering, University of Chinese Academy of Sciences, Beijing 100049, People's Republic of China

Abstract. Inspired by self-cleaning and water-repellent properties of the lotus leaf, biomimetic superhydrophobic cobalt blue/clay mineral hybrid pigments were facilely fabricated based on the rough surface of hybrid pigments and the modification with various organosilanes. The obtained hybrid pigments were characterized using various analytical techniques. Due to the difference in the compositions and morphologies of clay minerals, the obtained cobalt blue/clay mineral hybrid pigments exhibited different color properties. Superhydrophobicity of hybrid pigments was mainly regulated by the types of organosilanes instead of the morphologies of hybrid pigments. The sprayed coating of the superhydrophobic hybrid pigments exhibited the excellent self-cleaning performance with high water contact angle and low sliding angle. The coatings also presented excellent environmental and chemical durability even under harsh conditions. Therefore, the obtained biomimetic superhydrophobic cobalt blue/clay mineral hybrid pigments may be applied in various fields, such as anticorrosion, self-cleaning coating, etc.

Keywords: Biomimetic · Superhydrophobic · Cobalt blue · Clay minerals · Self-cleaning

1 Introduction

Cobalt blue (cobalt aluminate, $CoAl_2O_4$) pigment is a typical eco-friendly blue inorganic pigment. Due to high refractive index, excellent chemical and thermal stability, it can be widely applied in ceramics, paints, engineering plastics, etc. However, the high cost of $CoAl_2O_4$ pigment severely restrains their relevant applications due to the high price of cobalt compounds and the disadvantages of the traditional solid phase method (Armijo 1969). In addition, most of the common application fields of $CoAl_2O_4$ pigment are hydrophilic, which may be easily contaminated by dark liquid, oil stains, blot in our daily life. Therefore, it is very necessary to develop low-cost $CoAl_2O_4$ pigment with and self-cleaning ability.

Inspired by the unique water-repellent surfaces of the lotus leaf in the natural world (Barthlott and Neinhuis 1997), the design of superhydrophobic surfaces has become the focus in both fundamental research and industrial applications by construction of rough surface structure and modification using materials with low surface free energy. Recently, substrate-based inorganic hybrid pigments composed of an inorganic substrate coated with inorganic pigment nanoparticles have attracted increasingly attention. Our groups successfully prepared $CoAl_2O_4$ hybrid pigment after incorporation of different clay minerals (Mu et al. 2015; Zhang et al. 2017). Incorporation of clay minerals greatly decreased the production cost and calcining temperature for formation of spinel $CoAl_2O_4$, as well as preventing from the aggregation of $CoAl_2O_4$ nanoparticles after being uniformly anchored on the surface of clay minerals. Based on the rough surface of the hybrid pigments, it could realize the superhydrophobic modification of hybrid pigments using materials with low surface free energy. In this study, different $CoAl_2O_4$ hybrid pigments derived from kaoline (Kaol), palygorskite (Pal), halloysite (Hal) and montmorillonite (Mt) were prepared and modified using different organosilanes including octyl triethoxysilane (OTES), dodecyl trimethoxysilane (DTMS), hexadecyltriethoxysilane (HTES) and perfluoroctyl trimethoxysilane (PFOTMS), and the effect of the types organosilanes and the morphologies of hybrid pigments on the superhydrophobic properties was comparatively studied.

2 Methods and Approaches

Cobalt blue/hybrid pigments were prepared according to the similar procedure reported in our previous study (Zhang et al. 2017). Typically, 2.91 g of $Co(NO_3)_2{\cdot}6H_2O$, 7.50 g of $Al(NO_3)_3{\cdot}9H_2O$ and 1.09 g of clay minerals were added to 50 mL of water under magnetic stirring at 150 rpm for 30 min, and then 3 M NaOH was added dropwise into above mixture until the pH was reached to 10. The suspension was continuously stirred for 2 h at room temperature, and then the solid products were collected by centrifugation, washed with water and directly calcined at 1100 °C for 2 h with a rate of 10 °C/min. Next, the superhydrophobic modification of hybrid pigments was conducted in the ammonia saturated ethanol solution (Zhang et al. 2018). 0.54 g of organosilanes and 1.5 g of hybrid pigments with different weight ratios were firstly added into 45 mL above ethanol solution and stirred for 1 h at room temperature, and then 4.00 g of water was injected quickly into the solution and stirred for 24 h at room temperature. The solid products were finally washed with ethanol for three times and dried in an oven at 60 °C. The obtained samples were labeled as clay mineral-HP-organosilanes according to the involved clay minerals and organosilanes.

3 Results and Discussion

Table 1. Color parameters, water contact angle (CA) and sliding angle (SA) of superhydrophobic hybrid pigments before and after being treated at various conditions

Samples	Conditions	L^*	a^*	b^*	CA/°	SA/°
Kal-HP-HTES	-	38.1	2.6	−63.8	164.2	1.0
	98% H_2SO_4	37.4	2.6	−63.6	163.8	1.2
	3 M NaOH	38.1	2.6	−63.8	163.3	1.3
	UV for 3 days	38.0	2.6	−63.9	164.0	1.1
Pal-HP-HTES	-	16.2	−21.1	−23.5	148.1	2.4
	98% H_2SO_4	15.8	−20.8	−26.2	148.2	2.3
	3 M NaOH	16.7	−21.3	−24.8	147.5	2.4
	UV for 3 days	16.9	−22.2	−23.3	148.3	2.4
Hal-HP-HTES	-	55.8	−2.3	−55.0	151.2	1.2
	98% H_2SO_4	55.6	−2.4	−54.6	151.6	1.5
	3 M NaOH	55.8	−2.8	−53.7	151.8	1.8
	UV for 3 days	55.2	−2.6	−54.2	150.8	1.2
Mt-HP-HTES	-	24.5	−21.2	−32.3	156.4	3.7
	98% H_2SO_4	23.2	−21.3	−33.2	155.6	3.4
	3 M NaOH	24.6	−21.4	−32.7	157.5	3.1
	UV for 3 days	21.2	−20.4	−30.5	154.8	3.3
Kal-HP-OTES	-	38.1	2.5	−63.8	135.1	4.0
Kal-HP-DTES	-	38.1	2.6	−63.8	151.2	3.0
Kal-HP-PFOTMS	-	38.0	2.5	−63.7	165.2	2.0

The color parameters, CA and SA of superhydrophobic hybrid pigments before and after being treated at various conditions were summarized in Table 1. It was found that superhydrophobic hybrid pigments derived from different clay minerals presented different color properties (Fig. 1), which might be attributed to the difference in the compositions of clay minerals, especially Fe element. Among the employed clay minerals, the higher content of Fe element was observed in Pal and Mt than Kal and Hal, which decreased the color parameters of hybrid pigments. By contrast, hybrid pigment prepared using Hal exhibited the optimum color properties, and hybrid pigments obtained from Kal came second. In addition, the water contact angle and sliding angle of different hybrid pigments derived from different clay minerals had no obvious difference after being modified using HTES (Fig. 1). Except for Pal-HP-HTES, the values of CA and SA of them were higher than 150° and below 5°, respectively. Although the hybrid pigments were treated at various conditions, the color properties, CA and SA almost kept stable, indicating the excellent environmental and chemical durability. Furthermore, superhydrophobic Kal-HP modified with various organosilanes presented different CA and SA. With the increase in the carbon chain length, the values of SA increased while the SA values decreased. Meanwhile, the incorporation of fluorine atom in organosilanes also favored enhancing the superhydrophobic properties of hybrid pigments.

Fig. 1. Digital photos of (a) Hal-HP-HTES, (b) Mt-HP-HTES, (c) APT-HP-HTES and (d) Kal-HP-HTES

4 Conclusions

Superhydrophobic cobalt blue/clay mineral hybrid pigments with self-cleaning property and different colors were successfully prepared by modifying using organosilanes based on the rough surface. The color and superhydrophobicity were closely related to the compositions of clay minerals and the types of organosilanes.

Acknowledgements. The authors are grateful for financial support of the Major Projects of the National Natural Science Foundation of Gansu, China (18JR4RA001), and the Youth Innovation Promotion Association of CAS (2017458).

References

Armijo JS (1969) The kinetics and mechanism of solid-state spinel formation—a review and critique. Oxid Met 1:171–198

Barthlott W, Neinhuis C (1997) Purity of the sacred lotus, or escape from contamination in biological surfaces. Planta 202:1–8

Mu B, Wang Q, Wang AQ (2015) Effect of different clay minerals and calcination temperature on the morphology and color of clay/$CoAl_2O_4$ hybrid pigments. RSC Adv 5:102674–102681

Zhang AJ, Mu B, Luo ZH, Wang AQ (2017) Bright blue halloysite/$CoAl_2O_4$ hybrid pigments: preparation, characterization and application in water-based painting. Dyes Pigm 139:473–481

Zhang AJ, Mu B, Hui AP, Wang AQ (2018) A facile approach to fabricate bright blue heat-resisting paint with self-cleaning ability based on $CoAl_2O_4$/kaoline hybrid pigment. Appl Clay Sci 160:153–161

40

Fabrication of ZnO/Palygorskite Nanocomposites for Antibacterial Application

Y. Kang, A. Hui, and A. Wang(✉)

Key Laboratory of Clay Mineral Applied Research of Gansu Province, Center of Eco-Material and Green Chemistry, Lanzhou Institute of Chemical Physics, Chinese Academy of Sciences, Lanzhou, China

aqwang@licp.cas.cn

Abstract. ZnO/palygorskite nanocomposites were synthesized by chemical deposition and calcination process for antibacterial application. The results indicated that ZnO nanoparticles were deposited on the surface of rod-like palygorskite. Antibacterial evaluation confirmed that ZnO/PAL nanocomposites presented the good antibacterial behavior against *Escherichia coli* and *Staphylococcus aureus*, which was mainly attributed to the synergistic effect of ZnO and palygorskite.

Keywords: ZnO · Palygorskite · Nanocomposites · Antibacterial application

1 Introduction

Widespread overuse of the antibiotics on animals has been faulted for creating potentially threat for human beings (Li et al. 2014; Mckenna 2013). Antibiotic-resistant bacterial strains emerge and pose increasing health risks, and thus new antibacterials are urgently needed (Morrison et al. 2014). Inspired by the antibacterial property of nanocomposites in the nature, functional application of natural clay minerals are of great interest in academia and industry (Williams et al. 2011). It is well-known that zinc oxide (ZnO) possesses excellent antibacterial properties against gram-positive bacteria and gram-negative bacteria (Hui et al. 2016; Liu et al. 2019). However, it is difficult to prevent from the aggregation of ZnO nanoparticles during preparation process, which results in the adverse effects of the antibacterial properties. Palygorskite (PAL) as a kind of clay minerals is common in many parts of the world, typically forming in volcanic ash layers as rocks. Interestingly, PAL has one-dimensional rod like morphology, high specific surface area and better ion-exchange capacity, which can be adsorbed onto the bacterial cell by electrostatic adsorption. Therefore, it seems to be a promising attempt to achieve a possible synergistic effect by combine PAL and ZnO after loading ZnO nanoparticles on the surface of PAL. Therefore, ZnO/palygorskite (ZnO/PAL) nanocomposites were synthesized by chemical deposition and calcination process in this study, and the antibacterial properties against *Escherichia coli (E. coli)* and *Staphylococcus aureus (S. aureus)* were also investigated.

2 Methods and Approaches

Fabrication of ZnO/PAL Nanocomposites: In a typical synthesis of ZnO/PAL nanocomposites, 2 g PAL and 20 wt% $Zn(NO_3)_2 \cdot 6H_2O$ were dissolved into deionized water, and then 10 wt% NaOH solution was added into above solution for 2 h. The mixture was ultrasonically dispersed for 30 min and aged for 24 h. The powder was collected by centrifugation and dried at 80 °C for 6 h, and finally annealed at 400 °C for 3 h in muffle furnace.

Characterization: The products were characterized by X-ray diffractometer (XRD, D/MAX-2200, Rigaku, Japan), transmission electron microscopy (TEM, JEM-1200EX, FEI, USA) and energy dispersive spectroscopy (EDS) elemental composition analyzer. The specific surface area of the samples was evaluated by Brunauer-Emmett-Teller analysis (BET, Micromeritics, Norcross, USA).

Antibacterial Evaluation: *E. coli* and *S. aureus* were tested as a representative culture both Gram-negative and Gram-positive bacteria, which kindly provided by China Veterinary Culture Collection Center. The antibacterial activity of the sample was evaluated by examining the minimum inhibitory concentration (MIC).

3 Results and Discussion

The XRD patterns of PAL and ZnO/PAL nanocomposites were shown in Fig. 1. The diffraction peaks at 2θ of 8.38°, 13.74°, 16.34° and 34.38° were the characteristic peaks of PAL (Wang et al. 2015). The diffraction peaks of the obtained ZnO/PAL were corresponded to a wurtzite ZnO structure (JCPDS standard card 36-1451) (Hui et al. 2016).

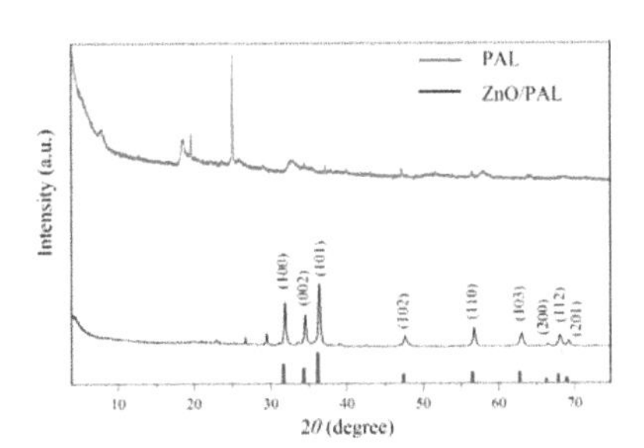

Fig. 1. XRD patterns of PAL and ZnO/PAL nanocomposites

The morphology of ZnO/PAL nanocomposites presented rod-like structure with a rough surface, and the length and the diameter were around 200 ~ 300 nm and a 30 nm, respectively (Fig. 2a), which indicated that ZnO nanoparticles were successfully loaded onto the rod-like PAL surface to form a heterostructure. The specific surface area of PAL and ZnO/PAL was found to be 173 $m^2 \cdot g^{-1}$ and 56 $m^2 \cdot g^{-1}$, respectively. Compared with pure ZnO, this strategy could obviously improve the specific surface area of ZnO, which was favorable to enhance the antibacterial activity of ZnO/PAL nanocomposites. The ring-type selected area electron diffraction pattern indicated the generated ZnO possessed polycrystalline nature (Fig. 2b). What's more, EDS result showed the surface element compositions of the as-prepared nanocomposites were Mg, Al, Si, Fe and Zn (Fig. 2c).

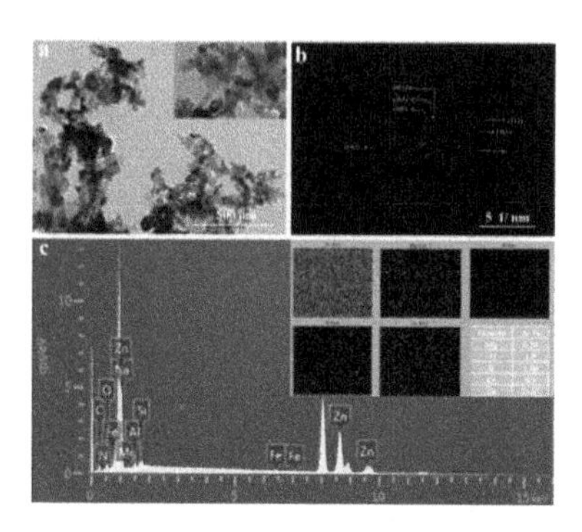

Fig. 2. (a) TEM image (inset is enlarged image), (b) selected area electron diffraction pattern and (c) EDS result of ZnO/PAL nanocomposites

As illustrated in Fig. 3, the microbial colonies of *E. coli* was visible when the concentration of sample was 0.25 mg·mL^{-1}, therefore, the MIC value of sample against *E. coli* was 0.5 mg·mL^{-1}. By contrast, there was small *S. aureus* colonies appeared in the plate when the sample contacts with *S. aureus* (Fig. 3i), and thus the MIC value of sample against *S. aureu* was 2.5 mg·mL^{-1}.

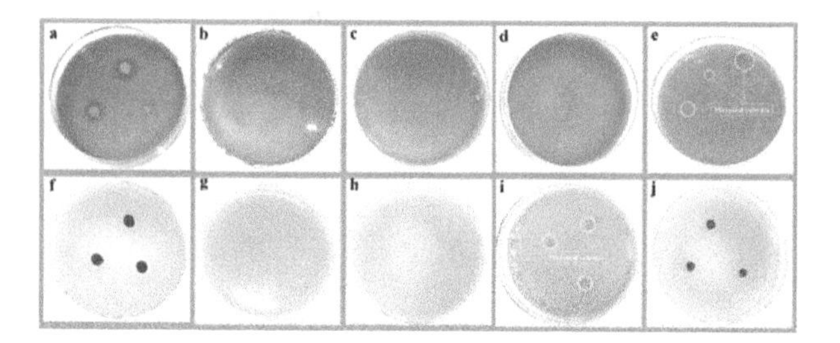

Fig. 3. (a, f) positive control of *E. coli* and *S. aureus*, *E. coli* treated by ZnO/PAL nanocomposites with various concentrations (b) 2.5 mg·mL^{-1}, (c) 1.5 mg·mL^{-1}, (d) 0.5 mg·mL^{-1}, (e) 0.25 mg·mL^{-1} and *S. aureus* (g) 5.0 mg·mL^{-1}, (h) 2.5 mg·mL^{-1}, (i) 1.0 mg·mL^{-1}, (j) 0.5 mg·mL^{-1}, respectively

In fact, PAL as a natural carrier with large special area, could absorb the bacterial cell by electrostatic adsorption. The reason for the synergistic antibacterial effect of ZnO/PAL nanocomposites was mainly due to the active factor of ZnO, which generated reactive oxygen species such as OH, O_2^1, O_2^{2-} and H_2O_2 (Hui et al. 2016; Liu et al. 2019; Ma et al. 2015). Incorporation of ZnO not only enhanced the antibacterial activity of natural PAL, but also reduced the cost of the preparation process, as well as efficiently realized the functional utilization of natural clay mineral resources.

4 Conclusions

In summary, ZnO/PAL nanocomposites were synthesized by chemical deposition and calcination process. The ZnO/PAL nanocomposites exhibited the excellent antibacterial activity against *E. coli* and *S. aureus*, and the MIC values for *E. coli* and *S. aureus* were 0.5 and 2.5 $mg{\cdot}mL^{-1}$, respectively.

Acknowledgements. This work was financially supported by the Major Projects of the National Natural Science Foundation of Gansu, China (18JR4RA001).

References

Hui AP, Liu JL, Ma JZ (2016) Synthesis and morphology-dependent antimicrobial activity of cerium doped flower-shaped ZnO crystallites under visible light irradiation. Colloid Surf A: Physicochem Eng Aspects 506:519–525

Li XN, Robinson SM, Gupta A, Saha K, Jiang ZW, Moyano DF, Sahar A, Riley MA, Rotello VM (2014) Functional gold nanoparticles as potent antimicrobial agents against multi-drug-resistant bacteria. ACS Nano 8:10682–10686

Liu JL, Wang YH, Ma JZ, Peng Y, Wang AQ (2019) A review on bidirectional analogies between the photocatalysis and antibacterial properties of ZnO. J Alloy Compd 783:898–918

Ma JZ, Hui AP, Liu JL, Bao Y (2015) Controllable synthesis of highly efficient antimicrobial agent-Fe doped sea urchin-like ZnO nanoparticles. Mater Lett 158:420–423

Mckenna M (2013) Antibiotic resistance: the last resort. Nature 499:394–396

Morrison KD, Underwood JC, Metge DW, Eberl DD, Williams LB (2014) Mineralogical variables that control the antibacterial effectiveness of a natural clay deposit. Environ Geochem Health 36:613–631

Wang WB, Zhang ZF, Tian GY, Wang AQ (2015) From nanorods of palygorskite to nanosheets of smectite via a one-step hydrothermal process. RSC Adv 5:58107–58115

Williams LB, Metge DW, Eberl DD, Harvey RW, Turner AG, Prapaipong P, Poret-Peterson AT (2011) What makes a natural clay antibacterial? Environ Sci Technol 45:3768–3773

Silicon Dioxide in Mineralized Heart Valves

A. Titov[1,2(✉)], V. Zaikovskii[1], and P. M. Larionov[1,3]

[1] National Research University, Novosibirsk, Russia
titov@igm.nsc.ru
[2] Sobolev V.S. Institute of Geology and Mineralogy of SB RAS, Novosibirsk, Russia
[3] Boreskov Institute of Catalysis of SB RAS, Novosibirsk, Russia

Abstract. This study showed that silicon dioxide of plant origin penetrated into the human body unchanged and was transferred through the blood to the heart, where absorbed by the pathological hydroxyapatite of mineralized heart valves.

Keywords: Silicon dioxide · Scanning and transmission electron microscopy · Mineralized heart valves · Bioavailability

1 Introduction

Over the past decades, numerous studies have shown that Si is an essential element and it affects human health. Silicon is present in the body as a trace element, but so far its biochemical function has not been confirmed by experimental data. Bioavailable silicon usually enters the body from solutions. Most easily it diffuses through the membranes and penetrates into the circulatory system in the form of orthosilicate acid, which is present in water, beer and some beverages. Among food products, the most significant sources of silicon are the products of plant origin: wheat, rice, oats, and barley. Despite the fact that vegetable food has a high content of silicon, its bioavailability is very limited due to the poor solubility of the forms of silicon present in the plants (Farooq 2015).

In this paper we present amorphous silica found in the composition of pathological formations - calcified heart valves. To explain the possible route of ingestion of amorphous silica, we considered one of the most common food crops for silicon consumption - rice.

2 Methods and Approaches

We studied intraoperative material, which included mineralized aortic and mitral valves of the heart, obtained from patients with acquired heart defects of rheumatic and septic genesis (Titov et al. 2016). Siliceous formations found in the calcifications of the heart valves were compared with the siliceous formations of plant origin: rice straw and husk.

The inventory of methods for structural and elemental analyses included high-resolution transmission electron microscopy (HR TEM), scanning electron microscopy (SEM), electron diffraction, and energy-dispersive X-ray spectroscopy (EDX). Electron microscopy was performed using a JEM2010 transmission electron microscope (acceleration voltage 200 kV, resolution 1.4 Å) equipped with an EDAX EDS detector (spectral resolution 130 eV) and a TESCAN MIRA3 scanning electron microscope with an Oxford EDS detector (resolution 128 eV) and built-in INKA ENERGY software.

3 Results and Discussion

Our investigation relied on the data obtained in the studies of the calcified formations on heart valves and their bioprostheses (Titov et al. 2016). The presence of silicon at the trace level was observed in the EDS analyzes of calcified heart valves. The studies of the dispersed material of heart valve calcifications by means of transmission electron microscopy revealed amorphous particles of rounded shape about 100 nm in diameter, along with hydroxyapatite nanocrystals (Fig. 1). The EDS spectra of these amorphous particles revealed Si, O, and C (carbon deposition), which corresponded to silicon dioxide (Fig. 1).

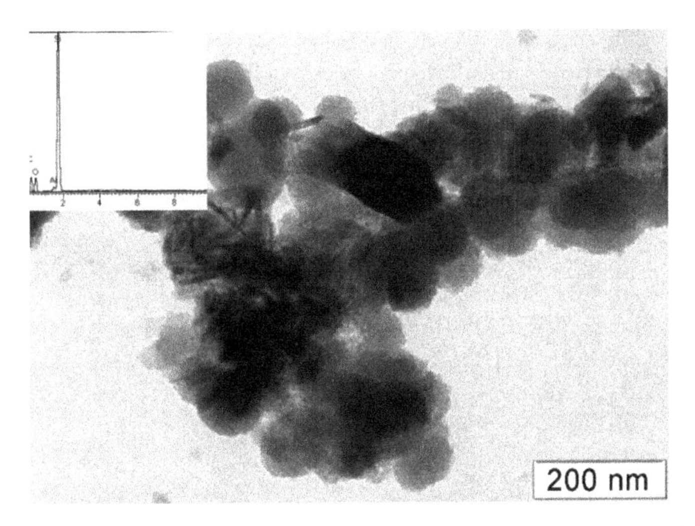

Fig. 1. Electron microscopic image (TEM) of amorphous silica particles among the dispersed calcified mineralized substance of a heart valve. Insert: EDX spectrum from one of the particles of silicon dioxide

As can be seen on the SEM images and EDS spectra of rice straw and husk (Fig. 2.), their surfaces were covered with a thin layer of silicon dioxide. After annealing the rice straw and husk at a temperature of 750 °C, white silica powder remained. The examination of the powder preparation by means of transmission electron microscopy showed that the rice substrate was represented by the rounded particles of amorphous silica with a diameter of about 50 nm (Fig. 3). The similarity of the amorphous particles of the rice substrate with the amorphous particles from heart valve calcifications in chemical composition and structure seemed quite obvious to us.

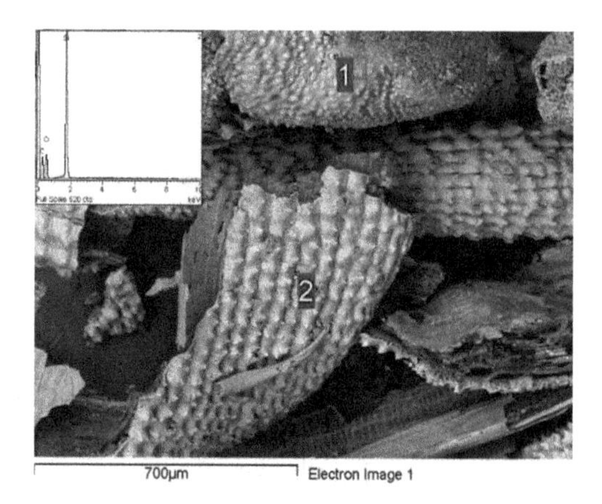

Fig. 2. SEM morphology of rice straw and husk. On the insert is the EDX spectrum from area 2 in the image.

Plants accumulate Si in the form of biogenic (phytolytic) silicon dioxide from soil solutions. It becomes included in plant tissues as a structural component imparting strength and rigidity to the stems. The main route of silicon intake into the organism starts from the gastrointestinal tract. Phytolytic silica is considered to be an insoluble form of Si.

However, most of Si is absorbed from solid products, therefore it is assumed that the phytolytic silicon dioxide is destroyed and absorbed (Jugdaohsingh 2007). Silicon dioxide detected by us in calcifications with nanocrystalline hydroxyapatite, remained unchanged. Apparently silica particles were transferred to the heart valves from the patients' blood and precipitated on hydroxyapatite. Hydroxyapatite is known to have a high sorption capacity (Titov et al. 2013). Perhaps a significant amount of silica can be deposited on the bone tissue as hydroxyapatite is one of its components. Our analysis of the chemical composition of calcified mineralized heart valves using an EDS spectrometer detected silicon at a trace level. The Si content in calcifications is quite significant for this trace element considering the sensitivity level of this method as 0.n weight percent.

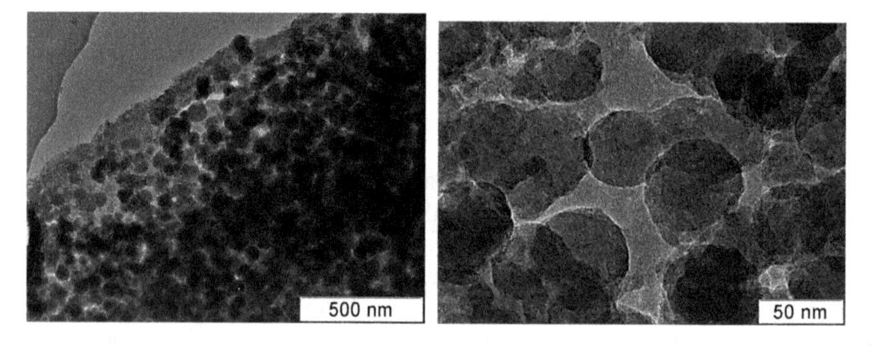

Fig. 3. A and B.Electron microscopic images (TEM) of cytolytic silica obtained from rice straw and husk after annealing at a temperature of 750 °C. B – the image of a fragment of A.

4 Conclusions

To conclude, we suggest that dispersed phytolytic silica may penetrate through the gastric tract into human blood in the unchanged form.

Acknowledgements. This research was carried out within the State Assignment to IGM SB RAS (project 0330-2016-0013).

References

Farooq MA, Dietz K-J (2015) Silicon as versatile player in plant and human biology: overlooked and poorly understood. Front Plant Sci 6:994–1023

Titov AT, Zaikovskii VI, Larionov PM (2016) Bone-like hydroxyapatite formation in human blood. Int J Environ Sci Educ 11(10):3971–3984

Jugdaohsingh R (2007) Silicon and bone health. J Nutr Health Aging 11(2):99–110

Titov AT, Larionov PM, Zaikovskii I (2013) Calcium phosphate mineralization of bacteria. In: Proceedings of the 11th International Congress For Applied Mineralogy (ICAM), pp 9–17

42

Microbial Colonies in Renal Stones

A. Izatulina[1(✉)], M. Zelenskaya[2], and O. Frank-Kamenetskaya[1]

[1] Department of Crystallography, St. Petersburg State University, St. Petersburg, Russia
alina.izatulina@mail.ru

[2] Department of Botany, St. Petersburg State University, St. Petersburg, Russia

Abstract. The presence and study of the species composition of bacterial and fungal colonies in renal stones was determined. It was shown that the presence of microorganisms depends on the phase composition of the renal stone. No microbial colonies were detected in oxalate stones. Under the action of the acid-producing bacterial and fungal colonies, secondary crystallization of calcium oxalates (whewellite and weddellite) on phosphate aggregates can occur.

Keywords: Renal stones · Calcium oxalates · Crystallization · Phosphate renal stones · Microorganisms

1 Introduction

Interest in pathogenic crystallization is growing every year, which is primarily due to the wide prevalence of diseases associated with stone formation, such as urolithiasis. Very few works are devoted to the influence of bacteria, viruses and micromycetes on stone formation in the human body. Thus, Sagorika et al. (2013) described a patient with renal aspergillosis along with urolithiasis, Zhao et al. (2014) reported on the crystallization of calcium oxalate in the presence of *E. coli*. Other studies have noted the initiation of calcium oxalate crystallization and aggregation in the presence of *E. coli*. Most of the known works are devoted to the so-called infectious renal stones, consisting mainly of struvite, and sometimes containing hydroxylapatite and brushite.

2 Methods and Approaches

The study was conducted using 21 samples of renal stones of different composition: 10 – oxalate, 5 – urate, and 6 – phosphate stones.

The substance of renal stones was sieved on the Czapek-Dox medium, potato glucose agar and Saburo medium to detect the presence and to determine the species composition of microorganisms. Preparation of nutrient media was carried out in accordance with GOST (GOST 9.048-89). The media was sterilized in an autoclave after preparation. After incubation period (20 days, 1 month), the samples were microscoped and examined in accordance with identifier (de Hoog and Guarro 1995).

Powder x-ray diffraction (PXRD) studies were carried out using the Rigaku «MiniFlex II» diffractometer (CuKα radiation of wavelength $\lambda = 1.54178$ Å, X-ray tube parameters were 30 kV/15 mA; highspeed solid state energy-dispersive detector LYNXEYE was used). X-ray diffraction patterns were collected at room temperature in the range of $2\theta = 5–50°$ with a step of $0.02°\ 2\theta$ and a counting time of half second per data point, the specimens were rotated 30 times per second during the data collection.

3 Results and Discussion

As the result of urate and oxalate renal stones sieving, the detection of micromycetes on the surface of the nutrient medium was random, and revealed a very small number of fungal colonies; the species of micromycetes were not constant. Abundant growth of fungal and bacterial colonies was detected on the phosphate renal stones: *Cladosporium cladosporioides, Penicillium expansum, Aspergillus niger*, sporiferous light colored fungus, *Geotrichum candidum, Candida sp., Fusarium chlamydosporum, Cladosporium sphaerospermum*, white and pink bacterial colonies (Fig. 1). Secretion of micromycetes on the surface of the nutrient medium was probably accidental, in many cases an insignificant number of fungal colonies was detected (1–3 colonies), the species distribution of micromycetes was not constant. There is a probability that some types of identified micromycetes (*Aspergillus niger, Cladosporium sphaerospermum, Phoma herbarum, Penicillium purpurogemum, Penicillium expansum, Fusarium chlamydosporum*) accidentally hit the test renal stone material (for instance, as a result of transportation or storage). The secretion of bacterial colonies and colonies of the fungus *Candida sp.*, is most likely not accidental, since a number of recent works indicate the possibility of a bacterial biofilm formation on the surface of renal (urinary) stones (Romanova et al. 2015). The possibility of micromycetes detection on the surface of renal stones within the microorganisms' biofilm requires further investigation.

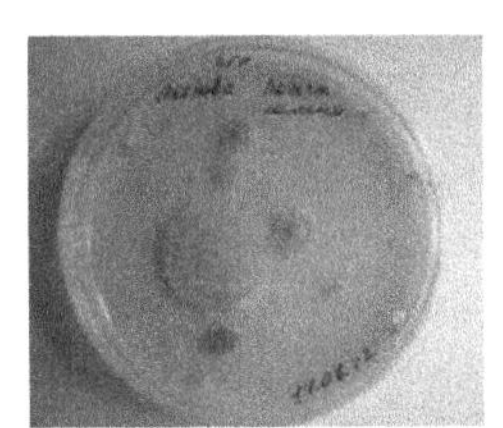

Fig. 1. Phosphate renal stone sample. The growth of bacterial and micromycetes colonies on the surface of the nutrient medium in a Petri dish (∅ 100 mm).

A colony of *Aspergillus niger* micromycetes was found on a nutrient medium on one of the phosphate samples, consisting of hydroxylapatite, struvite and brushite. Aspergillus infections have grown in importance in the last years (Hedayati et al. 2007). Oxalate crystals may be present in clinical samples, due to the high acid-producing ability of this fungus. When *Aspergillus niger* is growing on a liquid nutrient medium, it was found that acidification of the culture fluid begins almost immediately after spore germination and continues during the whole period of mycelium active growth. After a month of incubation (Fig. 2), under the influence of Aspergillus niger culture, the renal stone softens and many small crystals are observed during microscopy. According to the results of PXRD analysis, the observed crystals turned out to be mainly calcium oxalate dihydrate (weddellite); crystals of calcium oxalate monohydrate (whewellite) are also present but in subordinate quantities. Thus, under the influence of the *Aspergillus niger* culture, secondary crystallization of calcium oxalate occurs on a phosphate renal stone. The secondary crystallization of calcium oxalates under the action of microscopic fungi was recently described for monuments of cultural heritage (Rusakov et al. 2016). This phenomenon may be one of the reasons for the frequent presence of a phosphate nucleus in the center of an oxalate renal stones (Izatulina and Yelnikov 2008; Xie et al. 2014).

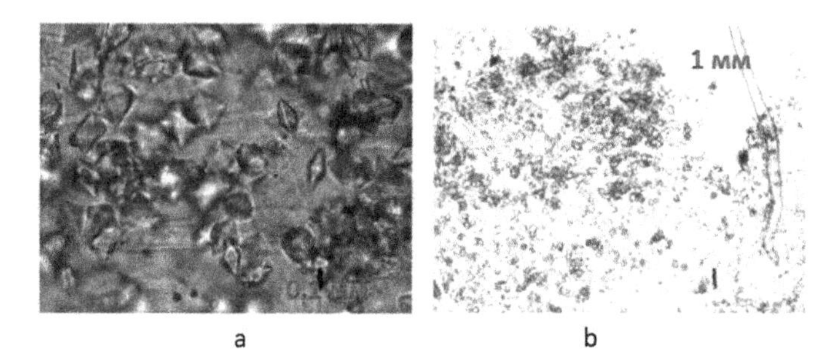

Fig. 2. The formation of calcium oxalate crystals in the presence of *Aspergillus niger* and a non-disrupting light-colored fungus during growth on a nutrient medium in a Petri dish (potato glucose agar was used as a nutrient medium).

4 Conclusions

Bacterial and fungal colonies were found on the surface of phosphate renal stones; no microbial colonies were found in oxalate and urate stones. It has been shown for the first time that under the influence of the microscopic fungus *Aspergillus niger*, secondary crystallization of calcium oxalates (whewellite and weddellite) can occur on phosphate aggregates. Thus, the possibility of the oxalate stones formation on phosphate nuclei with the participation of acid-producing bacterial and fungal colonies was shown.

Acknowledgements. This work was supported by the Russian Science Foundation (no. 18-77-00026). The XRD studies have been performed at the X-ray Diffraction Centre of St. Petersburg State University.

References

Bonaventura M, Gallo M, Cacchio P, Ercole C, Lepidi A (1999) Microbial formation of oxalate films on monument surfaces: bioprotection or biodeterioration? Geomicrobiology 16:55–64

De Hoog GS, Guarro J (1995) Atlas of clinical fungi, Baarn

Hedayati MT, Pasqualotto AC, Warn PA, Bowyer P, Denning DW (2007) Aspergillus flavus: human pathogen, allergen and mycotoxin. Microbiology 153:1677–1692

Izatulina AR, Yelnikov VY (2008) Structure, chemistry and crystallization conditions of calcium oxalates - the main components of kidney stones. In: Krivovichev SV (ed) Minerals as Advanced Materials I. Springer-Verlag, Heidelberg, pp 231–241

Romanova YuM, Mulabaev NS, Tolordava ER, Seregi AV, Seregin IV, Alexeeva NV, Stepanova TV, Levina GA, Barhatova OI, Gamova NA, Goncharova SA, Didenko LV, Rakovskaya IV (2015) Microbial communities on kidney stones. Molekulyarnaya Genetika, Mikrobiologiya i Virusologiya 33(2):20–25

Rusakov AV, Vlasov AD, Zelenskaya MS, Frank-Kamenetskaya OV (2016) The crystallization of calcium oxalate hydrates formed by interaction between microorganisms and minerals. In: Frank-Kamenetskaya OV, Panova EG, Vlasov DY (eds) Biogenic-Abiogenic Interactions in Natural and Anthropogenic Systems. Springer International Publishing, Switzerland, pp 357–377

Sagorika P, Viswajeet S, Satyanarayan S, Manish G (2013) Renal aspergillosis secondary to renal instrumentation in immunocompetent patient. BMJ Case Rep 2013:bcr2013200306

Xie B, Halter TJ, Borah BM, Nancollas GH (2014) Aggregation of calcium phosphate and oxalate phases in the formation of renal stones. Cryst Growth Des 15(6):3038–3045

Zhao Z, Xia Y, Xue J, Qingsheng Wu (2014) Role of E. coli-secretion and melamine in selective formation of CaC2O4·H2O and CaC2O4·2H2O crystals. Cryst Growth Des 14:450–458

43

Bacterial Oxidation of Pyrite Surface

S. Lipko[1(✉)], I. Lipko[2], K. Arsent'ev[2], and V. Tauson[1]

[1] Vinogradov Institute of Geochemistry SB RAS, Irkutsk, Russia
slipko@yandex.ru
[2] Limnological Institute SB RAS, Irkutsk, Russia

Abstract. The article considers the study of the role of bacteria in the surface oxidation of pyrite. The experiment provided the data on characteristic morphological changes of the surface and the first data on influence of a non-autonomous phase (NP) on bacterial oxidation.

Keywords: Pyrite · Iron-oxidizing bacteria · SEM-EDAX · Non-autonomous phase · Surface

1 Introduction

For many years, extensive studies have been conducted on the processes of diagenetic redistribution of ore-forming components in the Earth's lithosphere and the formation of iron-containing nodules and ores, aimed at the their prospective industrial use. Most researchers believe that the earliest microbial ecosystems were based on sulfur transformations – sulfate-reduction and disproportionation (Wacey et al. 2011).

The choice of pyrite as the object of study is geochemically justified by the close connections of iron sulfides with organic matter in various environments, including hydrothermal conditions (Lindgren et al. 2011).

Although the chemistry of the processes has been studied in principle, there remains a number of unresolved issues. The most important are proof of paleobacterial processes and determination of their role in the formation of mineral deposits. Here the range of opinions is very wide: from complete denial to recognition of their leading character at the sedimentary-hydrothermal stage of ore formation (Vinichenko 2007). The morphological effects of the interaction of mineral surfaces with bacteria have not been sufficiently investigated, which complicates interpretation of natural observations. In particular, it is unclear what effect non-autonomous phases located within the surface layer of the crystal have on the interaction of bacterial communities with the pyrite surface (Tauson et al. 2008, 2009a). Non-autonomous phases (NP) are nanocrystalline objects formed in the surface layer of the crystal through interaction with its surface of the growth medium components or contacting autonomous (classical) phases. The experiment within the framework of present research used specially synthesized pyrite crystals with different degrees of NP development on the surface (Tauson and Lipko 2013), with the aim to study the process of interaction between bacteria and NP and to establish the role of surface phases in the oxidative processes initiated by the acidophilic iron bacteria.

2 Methods and Approaches

The culture of acidophilic iron-oxidizing bacteria isolated from natural habitats (sulfide occurrences) of the Baikal area of the Irkutsk region was used for research on the bio-oxidation of the pyrite surface. This bacterial culture was provided by the laboratory № 7 of Irkutsk scientific-research Institute of rare and precious metals and diamonds, JSC "Irgiredmet". Iron-oxidizing bacteria are used in laboratory tests for bacterial oxidation of resistant iron-sulfide ores containing gold.

The synthesis of pyrite crystals was performed according to the standard technique of hydrothermal thermogradient synthesis in titanium inserts at T = 450 °C and 500 °C and a pressure of 1 kbar (Tauson et al. 2008). In the synthesis of pyrite $Fe^{+}S$ charge was used, the composition of the surface non-autonomous phase was regulated by the activity of sulfur depending on Fe/S ratio. The obtained crystals were up to 5 mm in size. Pyrite, obtained at high sulfur activity, contains virtually no NP on the surface. At lower sulfur activity, a layer of NP up to ~500 nm thick with a base composition similar to pyrrhotite, but with different forms of sulfur, is formed: Fe^{2+} $[S, S_2, Sn]^{2-}$ (Tauson et al. 2008). These surface formations are able to absorb cationic impurities and oxysulfide anions.

To conduct research on pyrite bio-oxidation, a mixture of acidophilic iron bacteria was grown on a liquid 9 K medium at room temperature and constant stirring within 5 days. In eight 250 ml conical flasks with 50 ml of 9 K medium (without $FeSO_4$) there were placed pieces of polished pyrite (4 flasks) and 5–6 pieces of pyrite with nonautonomous phases (4 flasks). The medium and pyrite flasks were sterilized at 0.5 atm. for 10 min to minimize pyrite oxidation. After cooling the medium, 6 flasks were inoculated with 5-day bacterial culture. Previously, the culture was centrifuged and washed from the medium residues with iron in 0.01 m H_2SO_4. The concentration of iron bacteria cells added to the medium with pyrite was about $1*10^7$ cells/ml. The remaining two flasks with polished pyrite and NP on the pyrite surface were used as a control, without bacteria. Cultivation took place at room temperature and constant stirring on a shaker (about 110 rotation/min) for three weeks. Every week 2 flasks with different samples of pyrite were selected for further research. The flasks with control samples were examined after 3 weeks of cultivation. The bacterial film from the pyrite samples were washed with 2% aqueous solution of Polysorbate Tween 80. Pyrite crystals washed after the experiment were dried in air and analyzed on the scanning multi-microscope SMM 2000 in atomic force mode, scanning electron microscope FEI Quanta Company (USA) 200 with energy dispersive device EDAX for X-ray microanalysis.

3 Results and Discussion

The experiment on pyrite bio-oxidation established that the surface of polished pyrite is less susceptible to bacterial oxidation as compared with NP-containing pyrite. Microphotographs show the surfaces of polished pyrite (roughness less than 5 nm) and NP-containing pyrite (roughness more than 300 nm) after two weeks of bacterial cultivation (Fig. 1).

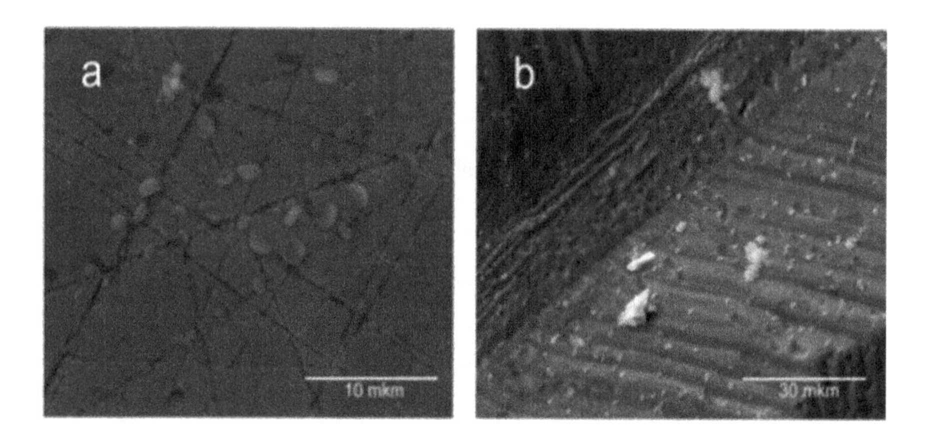

Fig. 1. Surface morphology of pyrite (scanning electron microscopy). a – polished surface, b – NP-containing surface.

The surface with NP exhibits bacteria and characteristic traces of interaction between bacteria and pyrite in the form of holes of different size comparable to the sizes of bacteria. For polished pyrite, these traces are almost absent and are observed only on the borders of scratches left from polishing. A similar result was obtained for the pyrite surface with minimal NP development synthesized at high sulfur activity. Therefore, the activity of bacteria is associated with the structure of pyrite surface. Similar formations, but significantly smaller (nano-holes) were discovered earlier, in the study of pyrites from the Sukhoi Log gold deposit (Irkutsk region) (Tauson et al. 2009b). This confirms the affinity of processes occurring in nature and in the experiment.

4 Conclusions

The research resulted in acquisition of data on pyrite bio-oxidation taking into account the structure of crystal surface under the given conditions. For this purpose, crystals with different degrees of non-autonomous phases development on the surface controlled by growth conditions were synthesized and used in the experiment for the first time. It was found that the surface of polished pyrite is less susceptible to bacterial oxidation, as compared with pyrite containing a non-autonomous phase. The resulting characteristic morphological changes in the surface will further be instrumental in addressing the issues of ore genesis, as well as identifying minerals that were formed at the initial or final stage of growth involving bacteria.

Acknowledgements. We thank Alexandra Mikhailova for suppling of bacterial culture. The research was performed within a state assignment, Project IX.125.3, No. 0350-2016-0025 and was funded by the Federal Agency for Scientific Organizations (FASO) within the framework of State Tasks No. 0345-2016-0003 (AAAAA16- 116122110061-6).

References

Lindgren P, Parnell J, Holm NG, Broman C (2011) A demonstration of the affinity between pyrite and organic matter in a hydrothermal setting. J Geochem Trans 12(3):3–7

Tauson VL, Lipko SV (2013) Pyrite as a concentrator of gold in laboratory and natural systems: a surface-related effect. In: Whitley N, Vinsen PT (eds) Pyrite: Synthesis, Characterization and Uses Chapter 1. Nova Science Publisher Inc., New York, pp 1–40

Tauson VL, Babkin DN, Lustenberg EE, Lipko SV, Parkhomenko IY (2008) Surface typochemistry of hydrothermal pyrite: electron spectroscopic and scanning probe microscopic data I. Synthetic pyrite. J Geochem Int 46(6):615–628

Tauson VL, Kravtsova RG, Grebenshchikova VI, Lustenberg EE, Lipko SV (2009a) Surface typochemistry of hydrothermal pyrite: electron spectroscopic and scanning probe microscopic data II. Natural pyrite. J Geochem Int 47(3):245–258

Tauson VL, Lipko SV, Shchegolkov YuV (2009b) Surface nanoscale relief of mineral crystals and its relation to non-autonomous phase formation. J Crystallogr Rep 54(7):1219–1227

Vinichenko PV (2007) Biogeology and Ore Formation. Izd-e Sosnovgeologiya, Irkutsk

Wacey D, Saunders M, Brasier MD, Kilburn MR (2011) Earliest microbially mediated pyrite oxidation in 3.4 billion-year-old sediments. J Earth Planet Sci Lett 301:393–402

Preparation of Macroporous Adsorbent Based on Montmorillonite Stabilized Pickering Medium Internal Phase Emulsions

F. Wang[1,2], Y. Zhu[1], W. Wang[1], and A. Wang[1(✉)]

[1] Key Laboratory of Clay Mineral Applied Research of Gansu Province, Center of Eco-Material and Green Chemistry, Lanzhou Institute of Chemical Physics, Chinese Academy of Sciences, Lanzhou, People's Republic of China
aqwang@licp.cas.cn

[2] College of Petroleum and Chemical Engineering, Qinzhou University, Qinzhou, People's Republic of China

Abstract. A macroporous material was prepared using oil-in-water Pickering medium internal phase emulsions (Pickering MIPEs) as template. The obtained macroporous materials with interconnected pore structure exhibited good adsorption capacities to Ce (III) and Gd (III) in water. The adsorption process could be achieved in 30 min, and the maximum adsorption capacities reached 230.64 mg/g for Ce (III) and 240.49 mg/g for Gd (III). Furthermore, the macroporous monolith exhibited excellent reuseability after consecutive adsorption-desorption cycles.

Keywords: Pickering emulsion · Medium internal phase emulsions · Rare metal · Adsorption · Porous material

1 Introduction

Compared with the conventional emulsion, Pickering emulsions exhibit peculiar long-term stability against droplet coalescence, in which the solid particles adsorbed at the oil-water interface act as a mechanical barrier to protect the dispersion phase liquid drops from the coalescence with the continuous phase (Tajik et al. 2017). Due to the unique properties, Pickering emulsions have been used as template to prepare porous polymer monoliths (Briggs et al. 2015). However, the dispersion phase of oil/water Pickering high internal phase emulsions mostly need a large amount of organic solvent, and thus it is indispensible to develop Pickering MIPEs by replacing poisonous organic solvent with low-cost and eco-friendly plant oil and reducing the internal phase volume.

Herein, monolithic macroporous materials were fabricated for adsorption of Ce (III) and Gd (III) based on Pickering MIPEs template, which was composed of the stabilizer of montmorillonite (Mt) and Tween-20 and the continuous phase of flaxseed oil. The effects of adsorption parameters including initial concentration and contact time on the adsorption properties were investigated, and the reusable performance of the adsorbent was also evaluated.

2 Methods and Approaches

Typically, macroporous carboxymethyl cellulose-g-poly(acrylamide)/montmorillonite (CMC-g-PAM/MMT) monolith was prepared based on Pickering MIPEs, which stabilized with 5% of Mt and 4% of Tween-20 (Wang et al. 2017). The obtained monolithic polymers were washed with acetone for 12 h and then immersed into 0.5 M NaOH aqueous alcohol solution ($V_{water}/V_{alcohol} = 3/7$) for 24 h to transfer the amide group to carboxyl.

The effect of the adsorption time and the initial concentration on the adsorption capacities were conducted according to the following procedure: 20 mg porous adsorbents were added into 25 mL Ce (III) and Gd (III) solution and shocked in a thermostatic shaker at 120 rpm and 30 °C for a given time. After the adsorption, the adsorbents were separated and the concentrations of the Ce (III) and Gd (III) were determined via UV-vis spectrophotometer using the chlorophosphonazo and azo arsine as the complexing agents, respectively. The adsorption capacities qe (mg/g) of the porous monolithic adsorbents were calculated according to the following equation:

$$q_e = \frac{(C_0 - C_e)V}{m} \tag{1}$$

where C_0 and Ce were the initial and equilibrium concentrations of Ce (III) and Gd (III) (mg/L), V (L) was the volume of the Ce(III) and Gd(III) solution.

The reusability studies were performed as follows: The adsorbents were desorbed by immersing 30 mL hydrochloric acid solution (0.5 M) for 2 h after adsorption, and then regenerated with 0.5 M NaOH solution. Finally, the adsorbents were filtered and washed to reach neutral using distilled water before next adsorption process. The adsorption-desorption cycle was repeated five times.

3 Results and Discussion

The representative images of macroporous monolith prepared by Pickering-MIPEs with 5% Mt and 4% Tween-20 as the stabilizer were shown in Fig. 1. The emulsions didn't flow in the inverted plastic centrifuge tube (Fig. 1a), indicating that the oil droplets were closely packed and the formed emulsion was a typical gel emulsion. The macroporous polymer monoliths of CMC-g-PAM/MMT were synthesized by free radical polymerization using APS as the initiator. The prepared wet monoliths were cut into pieces and Soxhlet extracted using acetone to remove the oil phase and surfactant, and then immerged into NaOH alcohol solution to complete hydrolysis of amide groups. Finally, the white monoliths of CMC-g-PAM/MMT were obtained after the dehydrated with acetone and dried in oven at 40 °C (Fig. 1b). The surface morphology of CMC-g-PAM/MMT was shown in Fig. 1c, and it presented a hierarchical pore structure with high connectivity. According to the statistical result of Image-Pro Plus 6.0 software, the average pore size of the macropore and the pore throat were 1.43 μm and 0.39 μm, respectively. Furthermore, the as-prepared porous materials exhibited narrow macropores and pore throats size distribution (Fig. 1d).

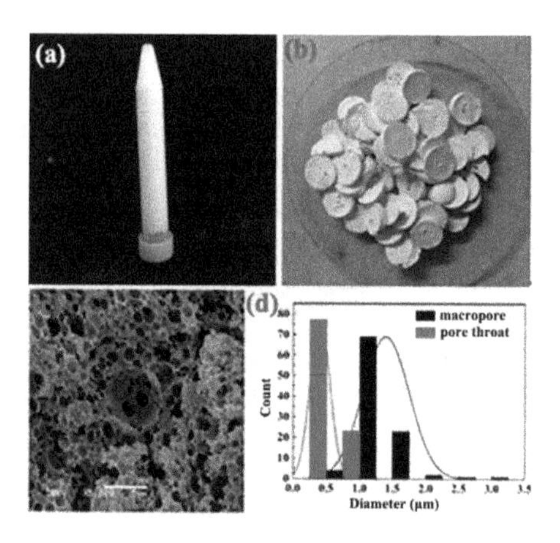

Fig. 1. Digital photographs of (a) the as-prepared Pickering MIPEs, (b) CMC-g-PAM/MMT monolith, (c) SEM image of porous CMC-g-PAM/MMT monolith and (d) pore size distribution of porous CMC-g-PAM/MMT

CMC-g-PAM/MMT monoliths were employed to remove of Ce (III) and Gd (III) from water. As shown in Fig. 2a and b, the adsorption capacities increased with the increase in the initial metal ions concentrations until the adsorption saturation was reached. The maximum adsorption capacities of the macroporous monoliths were 230.64 mg/g for Ce (III) and 240.39 mg/g for Gd (III). The higher adsorption capacity might be due to the sufficient functional groups and the highly interconnected pore structure. The effect of contact time of on the adsorption behavior was depicted in Fig. 2c and d. It was obvious that the porous monolithic adsorbent showed fast adsorption rate for Ce (III) and Gd (III), and the adsorption equilibrium could be reached within 30 min and 25 min Ce (III) and Gd (III), respectively. The macro-pores allowed fast and efficient mass transport, as well as provided sufficient contact between active groups and adsorbents, while the pore throats contributed to a high specific surface area.

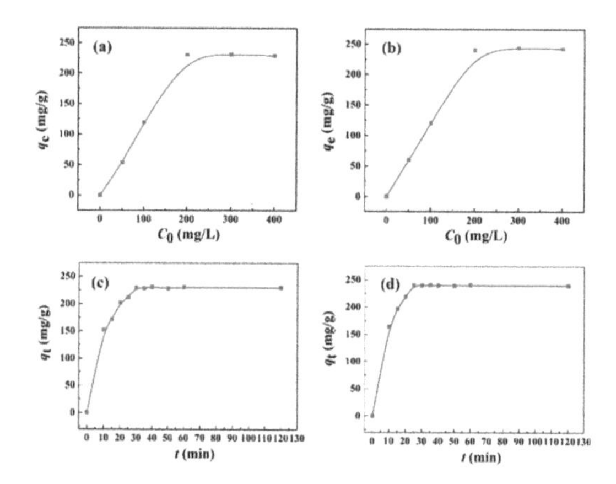

Fig. 2. Effect of the initial concentration of (a) Ce (III) and (b) Gd (III) on the adsorption capacity of porous monolith. Adsorption kinetic curves of the porous monolith for (c) Ce (III) and (d) Gd (III).

4 Conclusions

Macroporous polymer monoliths of CMC-g-PAM/MMT were successfully synthesized by free radical polymerization based on based on Pickering MIPEs stabilized with 5% of Mt and 4% of Tween-20. The as-prepared macroporous polymer monoliths possessed a hierarchical pore structure and highly interconnection, which favored enhancing the adsorption properties to Ce (III) and Gd (III), such as high adsorption capacity, quick adsorption rate, and good reusability.

Acknowledgements. The authors are grateful for financial support of the Major Projects of the National Natural Science Foundation of Gansu, China (18JR4RA001) and the National Natural Science Foundation of China (21706267).

References

Tajik S, Nasernejad B (2017) Surface modification of silica-graphene nanohybrid as a novel stabilizer for oil-water emulsion. Korean J Chem Eng 34:2488–2497

Briggs NM, Weston JS, Li B, Venkataramani D, Aichele CP, Harwell JH, Crossley SP (2015) Multiwalled carbon nanotubes at the interface of Pickering emulsions. Langmuir 31:13077–13084

Wang F, Zhu YF, Wang WB, Zong L, Lu TT, Wang AQ (2017) Fabrication of CMC-g-PAM/Pal superporous polymer monoliths via eco-friendly Pickering-MIPEs for superior adsorption of methyl violet and methylene blue. Front Chem 5:33

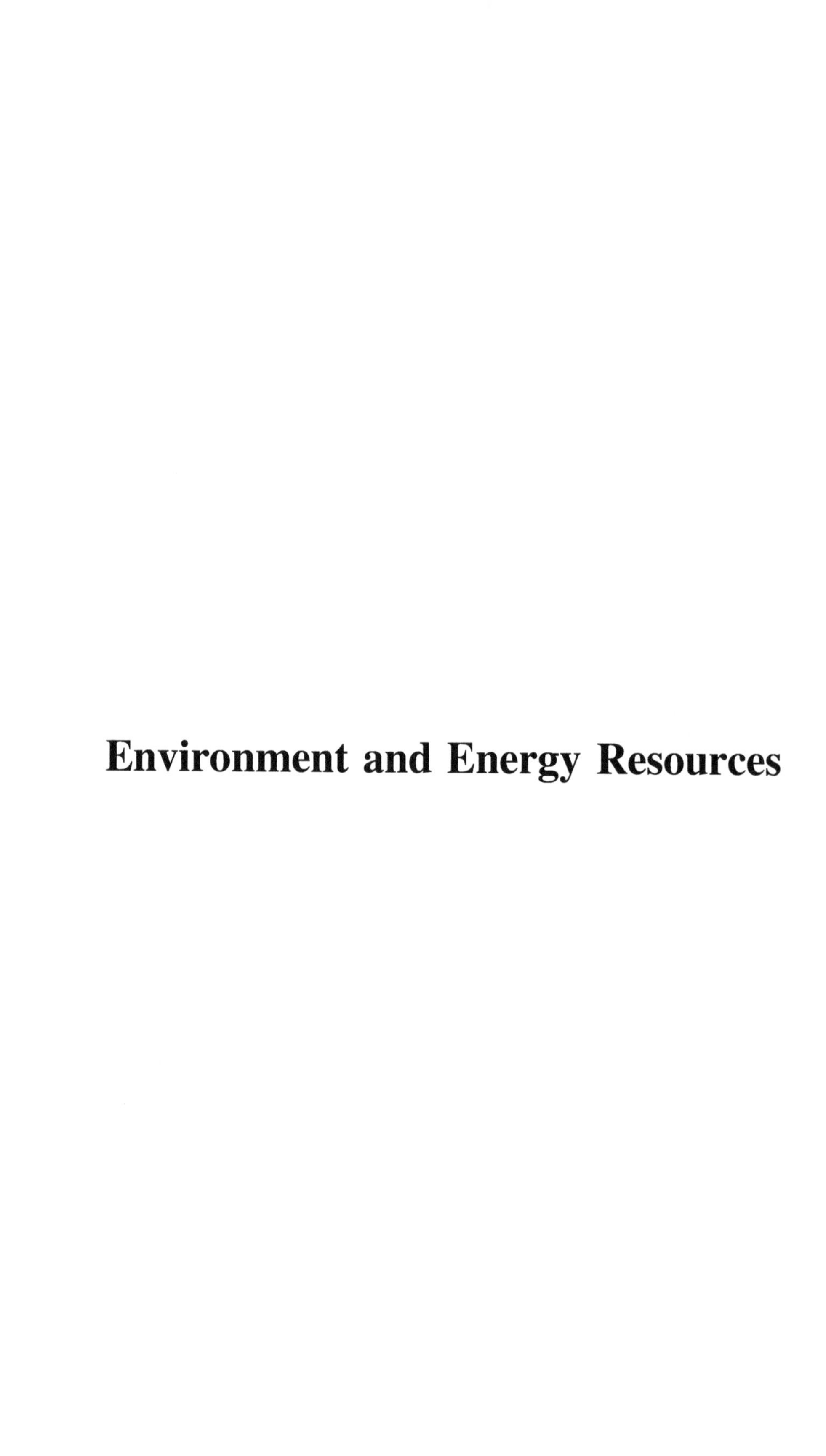

Environment and Energy Resources

Optimization of the Natural-Technical System "Iron Ore Quarry" Management Based on the Algorithm of the Rock Mass Stability Ensuring

L. Yarg(✉), I. Fomenko, and D. Gorobtsov

Department of Engineering Geology,
Russian State Geological Prospecting University (MGRI), Moscow, Russia
ifolga@gmail.com

Abstract. The method of natural-technical system (NTS) "Iron ore deposits" optimal control in terms of the pit walls stability is based on two-level systems with cross-links. The algorithm for optimizing the pit walls angles designed values includes the following steps: separation of rock massif into engineering-geological complexes (EGC), typing of the pit walls within the EGC, substantiation of the calculation geomechanical models and stability analysis of the pit walls based on mathematical modeling. Based on the results of the calculations the maximum angle of the pit wall is determined at which it remains stable. As minimized performance criteria the deviations of the stability factors current state from the maximum allowable values are considered. The proposed approach is one of the ways to ensure the stability of the deep-pit quarries walls during their long-term development.

Keywords: Natural-technical system (NTS) · Open pit · Stability assessment · Optimization of pit walls angles · Control of NTS

1 Introduction

The development of iron ore deposits and permanent deepening of the open pit leads to changes in the stress state, decompaction of rocks, an increase in massif fracture, weathering rates and a decrease in the strength properties of rocks that form the open pit, activation of geological processes: debris, rock falls, landslips and landslides, suffusion, surface erosion.

The considered natural-technical system "Ore deposits of KMA" is a complex system of the local level. The functioning of the local NTS "Iron Ore Quarry" is characterized by: a certain set of processes developing permanently without the stabilization stage under the influence of long-term man-made interactions which form the basis of the NTS operation. Reduction of negative consequences is possible only with a clear understanding of the processes developing in the field of interaction of natural-technical systems (NTS) "mining and processing plant (GOK") (Yarg et al. 2018).

Research objective: optimization of the NTS "iron ore quarry" management based on the algorithm of the rock mass stability ensuring.

2 Methods and Approaches

The processes development initiated by technological work is progressive in space and time. Long-term exploration of deposits leads to the changes of boundaries, mode and set of processes (Bondarik and Yarg 2015).

The system of engineering geological support in the quarry areas includes a range of work and research aimed to obtain the information about engineering geological conditions during the entire life of the quarry, assessment and forecast of the slope stability at various stages of their construction to achieve the technical, economic and environmental safety of mining work.

Effective management of the natural-technical system "Iron Ore Deposit KMA" should be carried out taking into account both local and global stability factors of the pit walls.

Separation of the Rock Massif to EGC. Features of engineering-geological conditions including lithologic-petrographic composition, physical and mechanical properties, structural disturbance, parameters of the natural stress field require an individual approach to the process of predicting the behavior of an array of rocks. This becomes possible only on the basis of correct engineering and geological research data.

Elementary NTS "Stoilensky Quarry" is divided into two engineering-geological complexes (Yarg et al. 2018):

- The upper one is composed of loose and semi-rock soils with a thickness up to 90 m. The sedimentary cover is typified taking into account the geological structure, hydrogeological conditions (the water inflow along the open-pit contour water permeability) and the physical and mechanical properties of the soils.
- The lower EGC is represented by rocks with a thickness of up to 600 m. The main stability determining factors are: anisotropy of the massif properties due to its fracturing and spatial orientation of the cracks.

Engineering geological processes developing during the operation of the elementary NTS "Open pit" of the iron ore deposits of Stoilensky and Lebedinsky GOK are: scree formation; collapse; landslides; surface erosion; suffusion: mechanical, chemical; filtration deformations.

Tiping of the Pit Walls Within the EGC. A "bowl" of a quarry with a simple structural plan of the rock mass (Fomenko et al. 2016, Hoek and Bray 1981, Wyllie and Mah 2010) (i.e. assuming that the direction of weak zones and fracturing within the pit remains constant) can be divided into zones of conditional stability and potential instability of the walls (Fomenko et al. 2016).

In accordance with these factors, three types of pit walls quarrying were identified: relatively difficult, difficult and very difficult.

3 Results and Discussion

Optimization of the NTS "Mineral Deposit" functioning is based on a modern methodology for stability calculation (Pendin and Fomenko 2015, Bar et al. 2018).

For potentially unstable pit wall the probable collapse can occur according to the following schemes:

1. The azimuth of crack systems fall coincides with the azimuth of the pit wall fall. In this case a flat problem can be solved.
2. The azimuth of crack systems fall does not coincide with the azimuth of the pit wall fall, but at the same time according to the kinematic analysis results the formation of wedge-type collapses is likely. In this case the pit wall stability problem is solved in a three-dimensional formulation, for example using the method of volume blocks.

Based on the results of the calculations the maximum pit wall angle is determined at which it remains stable.

In accordance with the "large-scale interconnected" theory (Tsurkov and Litvinchev 1994), the management of local NTS "Iron Ore Quarry" in terms of the pit walls stability can be based on two-level systems with cross-links. As minimized performance criteria the deviations of the stability factors current state from the maximum allowable values are considered. As optimized parameters the following were taken: the level of the upper Jurassic aquifer, the strength properties of Alb-Cenomanian sands and Devonian clays, fracturing and blockiness of the Precambrian massif.

The graphs (Figs. 1, 2 and 3) of the relation between safety factor and the dynamics of aquifer, blockiness and strength properties of rocks allow setting the limit values of the system coordinates at which the system does not leave the zone of admissible states.

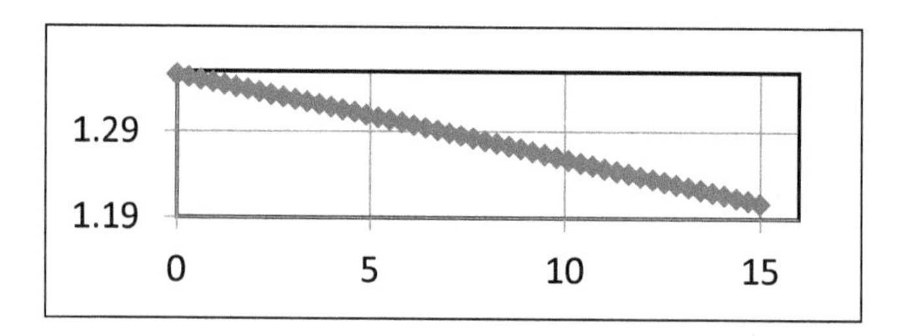

Fig. 1. The effect of groundwater level rise (horizontal axis) on the global safety factor of the pit walls (vertical axis).

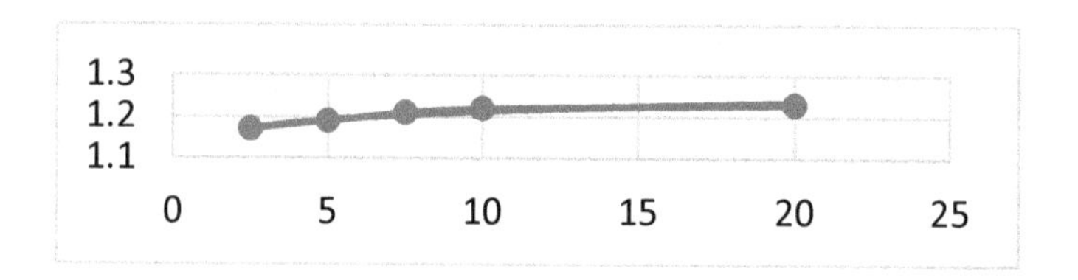

Fig. 2. The relation between Ku (vertical axis) and the blockiness of the rock massif (distance between cracks), (horizontal axis)

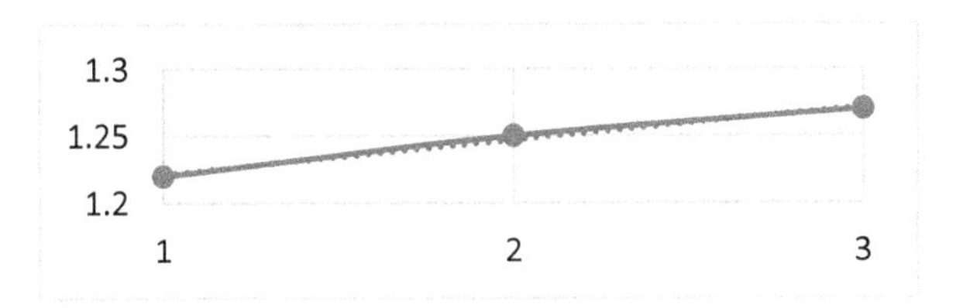

Fig. 3. The effect of the ore-crystalline rocks strength (horizontal axis) in the massif on the safety factor (vertical axis): 1 – densely fractured rocks; adhesion C = 690 kPa, angle of internal friction φ = 32°; 2 — moderately fractured, adhesion C = 1130 kPa, angle of internal friction φ = 36°; 3 — weakly fractured, adhesion C = 3140 kPa, angle of internal friction φ = 39°

4 Conclusions

The obtained results are in good agreement with the generally accepted ideas about the causes of stability infraction in the massif zones near the pit walls. At the same time the obtained data advantage is the possibility of establishing the limiting values of changes in the safety factors on graphs. These safety factors determine both local and global stability of the pit walls at which the system will not leave the zone of admissible states.

Control solutions that ensure the safe work performance should include: adjustment of the drainage system and water supply process, taking into account the position of the GWL; interception and organized disposal of surface and seepage water into the circulating system; maintaining the water level in the settling ponds, excluding flooding of the territory; adjustment of the blasting operations technology during the quarrying of deep horizons taking into account the stress state of the massif and the occurrence of the rock technogenic fracturing during drilling and blasting operations.

The obtained predictive estimates of the pit walls stability can be used in the design and development of fields with similar engineering geological conditions.

References

Bar N, Weekes G, Welideniya S (2018). Benefits and limitations of applying directional shear strengths in 2D and 3D limit equilibrium models to predict slope stability in highly anisotropic rock masses. https://www.researchgate.net/profile/Neil_Bar/research

Bondarik GK, Yarg LA (2015) Engineering geology. Questions of theory and practice. Philosophical and methodological foundations of geology. KDU (in Russian)

Fomenko IK, Pendin VV, Gorobtsov DN (2016) Estimation of the stability of quarries of quarries in rocky soils. Min Sci Technol (3), 10–21 (in Russian)

Hoek E, Bray JW (1981) Rock slope engineering, 3rd edn. Institution of Mining and Metallurgy, London

Pendin VV, Fomenko IK (2015) Methodology of landslide hazard assessment and forecast. Publishing House of the Russian Federation Lenand, Moscow (in Russian)

Tsurkov VI, Litvinchev IS (1994) Decomposition in dynamic problems with cross-links. Science: Physics and Mathematics, Moscow (in Russian)

Wyllie DC, Mah CW (2010) Rock slope engineering: civil and mining, 4rd edn. Spon

Press/Taylor&Francis Group, London
Yarg LA, Fomenko IK, Zhitinskaya OM (2018) Evaluation of slope optimization factors for long-term operating open pit mines (in terms of the Stoilensky iron ore deposit of the Kursk Magnetic Anomaly). Gornyi Zhurnal (11), 76–81

46

Environmental Pollution Problems in the Mining Regions of Russia

E. Levchenko(✉), I. Spiridonov, and D. Klyucharev

FSBI IMGRE, Moscow, Russia
imgre@imgre.ru

Abstract. The main types of environmental impact during exploration, development and mining of mineral deposits are considered. The indicators of the environmental situation caused by the mining and mineral processing in the mining regions, as well as the environmental consequences of accumulated mining and industrial waste are presented. The results of environmental monitoring of the Russian industrial cities are demonstrated.

Keywords: Environmental safety · Mining · Man-made waste · Heavy metals · Pollution of the ecosystem

1 Introduction

Intensive economic development due to the steady progress in science and technology entails an inevitable increase in the consumption of minerals. In this regard, the increase in mineral production during the last century, a sharp increase in the mining activities contributed to the accumulation of mining waste and man-made pollution of ecosystems. Besides, despite the obvious benefits of mining for the benefit of man, on the other hand, it is also a powerful source of environmental hazards for biota and humans (Aleksandrova and Nikolaeva 2015).

Many chemical elements contained in waste products, in addition to industrial value, cause toxic effects on the ecosystem.

The mining of mineral deposits leads to a change in the basic physicochemical properties of the lithosphere, including its main functions, i.e., geodynamic, geophysical, resource, and geochemical. The study of changes in the ecosystem's parameters during the life period of a mining enterprise is one of the key goals of an ecological-geochemical assessment.

The high level of the environmental impact is typical of the waste produced by ore processing and metallurgical operations, since their storage requires special engineering structures, and the waste contains chemical components harmful to nature and human health. Their mass is inferior to that of stripped overburden and host rocks, but they affect the environment more perniciously (Spiridonov and Levchenko 2018).

The environmental situation has deteriorated significantly due to the fact that at the end of the last century after the collapse of the USSR, many large mining complexes did not cope with economic difficulties and ceased their activities. The tailings of the enterprises, by majority toxic, have remained uncontrolled. Their conservation and reclamation have not been carried out timely; hence pollution keeps on growing. In the soils buried under the dumps tangible geochemical transformations occur. The soils buried 20 and more years ago display a strong oxidation over the whole depth of their profile (e.g., pH stays as low as 3.5–4.0), and soil colloids become destroyed. The soil absorbing complex is disturbed, the mobility of organic matter increases, the soil horizons gain ore components, which additionally differentiate due to unequal mobility. These facts testify the mobility of chemical elements in the dumps, and the latter often remain connected to the watercourse systems and can affect the territory of the mining and processing works in the area of air emissions and waste storages.

2 Methods and Approaches

Monitoring of the natural environment should be carried out at all stages of the mining area life, from exploration to mining and further reclamation of disturbed lands and until the site becomes completely stabilized.

The basis of this paper are ecological and geochemical studies, including the identification of areas of environmental pollution by toxic substances, assessment of their extent and composition of their pollution; assessment of potential geochemical endemicity; zoning of the territory according to the pollution level and the degree of environmental danger. identification of pollution sources; identifying areas of potential man-made objects; ecological and geochemical monitoring and forecast of the development of negative processes; development of recommendations for the rehabilitation of areas of poor ecological condition; identification of populations with an increased risk of morbidity. The result of these studies is the compilation of ecological and geochemical maps portraying the ecological status of the territory.

The study of the environmental health is carried out in the following main areas: mapping of the man-made pollution in soil and snow cover; establishing the characteristics of the response of plants to soil pollution; geochemical studies of ground and surface water, and stream sediments; analysis of the chemical composition of atmospheric air, precipitation and aerosols, industrial waste materials as sources of environmental pollution and objects for the extraction of secondary raw materials; relationships of environmental pollution and health indicators of the population living in the pollution hot spots.

3 Results and Discussion

The share of mining industries accounts for 70–80% of the volume of all man-made formations, which have their own characteristics, due to the composition of the feedstock, the technology of extraction, enrichment or processing, and a number of other factors.

As demonstrated by ecological and geochemical studies, the most serious negative effects are related to: the functioning of large industrial hubs (Nizhny Novgorod, Irkutsk-Cheremkhovo, Khabarovsk, Vladivostok, etc.), as well as exploration and development of mineral deposits in active mining areas (Kirovsk, Mama-Bodaibo, Khapcheranga, Dalnegorsk-Kavalerovsk, Norilsk, and other areas of similar profile).

On the basis of the analysis of the updated database of available technogenic objects, including rare metal deposits, the allocation of 576 technogenic formations on the territory of the Russian Federation is analyzed.

Relevant location maps were compiled, and ranking of technogenic deposits and formations was carried out using the following parameters: areal extent, storage type(s), type(s) of technogenic formations, hazardousness classes, and level of environmental impact. The man-made deposits and formations were ranked by their effect on the elements of the environment.

The analysis of the hottest spots suggests that a series of the factors provokes the deterioration of the ecological situation in the territories.

Of particular concern is the ore processing plant waste, since it requires special engineering structures, and the waste itself contains chemical elements and compounds harmful to nature and human health. Their amounts are inferior relatively to the masses of stripped barren overburden and hosting rocks, but they affect the ecological situation more perniciously. For example, the environmental situation caused by the extraction of mineral raw materials and the disposal of waste on more than 25% of the territory of the Urals economic region is estimated as a crisis. Slightly less than the area of such lands in the south of the Russian Far East, Khanty-Mansi Autonomous Area, Tyumen Region, Krasnoyarsk Territory and other areas of intensive mining and processing of mineral resources.

According to the environment impact degree, the highly hazardous objects list is as follows: apatite concentrates of the Khibiny apatite-nepheline deposits (TR, Sr, F), enrichment tails of the eudialyte lujavrites of the Lovozero GOK (TR, Th), tailings of enrichment of baddeleyite-apatite-magnetite ores of the Kovdor Mining apatite, baddeleyite ZrO_2). Medium-level objects are waste storages accumulated from the apatite concentrate processing in the Khibiny group deposits (phosphogypsum) containing rare earth metals and gypsum (Bykhovskiy et al. 2016; Karnachev et al. 2011).

The toxicity of mining products depends on their physical condition and chemistries. Understanding the mechanisms of the action of chemical elements and compounds on the environment and public health makes it possible to optimize medical consequences and to carry out acceptable mining and processing of mineral raw materials. At the same time, it is necessary to take into account the whole range of sources and objects of impact in order to create a system of medical and environmental safety of the work areas.

The problems of the urbanized environment as a human habitat become similar to those experienced by geologists, representatives of related professions and the population of geological exploration, mining, oil and gas, and metallurgical enterprises.

Three indicators are accepted in Russia as measures of the soil chemical pollution in Russia; these are the maximum acceptable content (MAC), the background content (Z_b) and crustal abundance/clarke (Zc). We analyzed the weighted average bulk content distribution of heavy metals (the hazardousness classes 1 and 2): Pb, Cd, Hg, Zn, Ni, and Cu. By the above mentioned three evaluation criteria, the cities falling into the 1st (highly dangerous) category are Irkutsk, Penza, Saratov, Chelyabinsk, Yekaterinburg, the 2nd (dangerous) are Perm and St. Petersburg, and the 3rd (moderately dangerous) include Blagoveshchensk and Vologda.

4 Conclusions

The extent of the loss of land, water, forest, recreational and other resources from subsoil use in general and from unused waste in particular places these processes on a par with negative factors that pose a threat to the country's security.

The environmental consequences of accumulated mining and industrial waste are larger than it is declared in various publications concerning the problem under consideration and are of a global scale.

References

Aleksandrova TN, Nikolaeva NV (2015) Ecological-geochemical estimate of the Russian mining and metallurgy waste. Polytech University Publishers, St. Petersburg (in Russian)

Bykhovskiy LZ, Potanin SD, Kotelnikov EI, Anufrieva SI et al (2016) Rare earths and Sc-bearing man-made formations and deposits in Russia. In: Rare earth and Sc minerals in Russia: Mineral commodities, VIMS Economic Geology Series, No. 31, pp 112–120 (in Russian)

Karnachev IP, Zhirov VK et al (2011) Ecological and sanitary estimate of the Khibiny mining area, Murmansk oblast. Vestnik MGU, vol 14, no 3, pp 552–560 (in Russian, with English abstract)

Spiridonov IG, Levchenko EN (2018) Mining waste and ecological safety. Prospect and protection of mineral resources, no 10, pp 15–24 (in Russian, with English abstract)

47

Depletion of the Land Resources and its Effect on the Environment

M. Abou Zahr Diaz, M. A. Alawiyeh, and M. Ghaboura(✉)
Department of Mineral Developing and Oil & Gas Engineering, Engineering Academy, RUDN University, Moscow, Russia
mostafa_ghab@live.com

Abstract. Resources depletion refers to the situation where the consumption of natural resources is faster than it can be replenished. In order to achieve economic growth, developing countries are abusing their lands on the grounds of economic interests. Population Explosion is acting as a catalyst for resources depletion. It seems evident that developing countries pursuing rapid economic growth disregard environmental concerns. The natural resources contribute at large to the economic development of a nation. Consumption pattern if not addressed will lead to irreversible climate change and declined economic growth, as a result of increased social, economic, and environmental costs and decreased productivity. Resource utilization has always been part of human history; however, the acceleration of economic growth activities together with the pursuit of an urgent economic development is the core cause of resources overexploitation. Consumption pattern will lead to irreversible climate change and declined economic growth.

Keywords: Resources · Economics · Population · Utilization · Depletion of the soil

1 Introduction

In addition to the problems of each civilization, humanity is faced with the urgent need to solve planetary problems. At the end of XX century, the first terrible signs of the deterioration of the quality of the biosphere had already appeared as a result of the development of man-made civilization and the installation to conquer nature started.

Smog over large cities, deforestation and the onset of deserts, depletion of the soil and basins of many rivers, a decrease in the number of fish and wild animals – all this worried people at the beginning of the twentieth century.

2 Methods and Approaches

A no less formidable problem is the ecological catastrophe approaching the planet. At present, mankind produces organic waste in an amount of two thousand times more than the waste volumes of the rest of the biosphere. Obviously, the violation of this equilibrium caused a whole complex of complex problems. Man, unlike all living things, is not strictly bound by the environmentally friendly conditions of his being, in a certain sense he was always going against nature, not adapting to it, but changing it in accordance with its needs (Satterthwaite 2009).

3 Results and Discussion

The demographic problem has become global long ago. In 1987, the five billionth inhabitant of the planet was born, and the growth rate is now such that every second the number of people on Earth is increased by three people (Anderson 2012). According to the figurative expression of scientists, the Earth is now "biting man" and it is quite natural to expect a demographic collapse in the near future, that is, a sufficiently sharp decline in population.

It can be caused by global hunger, depletion of mineral resources and soil, poor drinking water, thermal overheating of the surface, etc.

4 Conclusions

The current generation cannot help thinking about future children and grandchildren, who are to continue to carry the baton of history. Unfortunately, our civilization largely lives at the expense of the future, exhausting irreplaceable resources (oil, gas), polluting water, air and soil with its imperfect technologies, preserving many archaic social structures, sowing seeds of national and religious hatred that will sprout another century.

Based on the above, it can be concluded that only the efforts of the entire world community can prevent an environmental catastrophe that threatens all life on Earth.

References

Anderson R (2012) Resource depletion: opportunity or looming catastrophe? https://www.bbc.com/news/business-16391040

Satterthwaite D (2009) The implications of population growth and urbanization for climate change. Environ Urbanization 21(2):545–567

48

Calcite Mineral Generation in Cold-Water Travertine Huanglong, China

F. Wang, F. Dong(✉), X. Zhao, Q. Dai, Q. Li, Y. Luo, and S. Deng

School of Environment and Resource,
Southwest University of Science and Technology, Mianyang, China
fqdong@swust.edu.cn

Abstract. Mineral generations could help us to understand the physical, chemical and biological processes within their formation, and then to reconstruct the sedimentary paleo-environment and paleo-climate. The calcite in the Huanglong cold-water travertine can be divided into three mineral generations, which reveal two different sedimentary environment systems respectively. In the calcium cycle, calcite mineral generation exposes a step in recycling marine matter to the land, and it also allows the land to proliferate, which mainly manifeste in the addition of plant debris, algae and microbial residues, so that the topography has been accumulating.

Keywords: Cold-water travertine · Mineral generation · Paleo-environment · Huanglong

1 Introduction

Calcite is the main mineral component of travertine/tufa, and it plays a decisive role in the sedimentary evolution of travertine, whether inorganic or bio-organic (Pentecost 1995). The size of the calcite in travertine is a reflection of the deposition rate and can therefore be used to characterize its sedimentary environment, which is the result of physical, chemical and biological synergy during the deposition process. Herein, we divide the calcite in the Huanglong cold-water travertine into different mineral generations according to the sedimentary environment and evolution time series, i.e., from the generation of the parents to the descendants. The classification of these mineral generations helps to understand the physical, chemical and biological processes within their formation, and then to reconstruct the sedimentary paleo-environment and paleo-climate. On the other hand, the mineral generation of calcite will help to understand the architecture of travertine landscape (Wang et al. 2018), so that they can be better protected and leave more natural heritage of travertine for human beings.

2 Methods and Approaches

A comprehensive field geological survey of rocks consisting of calcite was performed, mainly from sedimentary rocks, and systematic sample collection based on the geological background of these rocks was carried out. The mineralogy studies of calcite were carried out by polarized light microscopy, XRD and SEM to determine their generational relationship.

3 Results and Discussion

From the diagenetic time series of calcite, the types of rocks are Mesozoic limestone and dolomite, and the travertine deposited since the Late Cenozoic. The calcite in travertine is further divided into two mineral generations, namely calcite and secondary in primary travertine travertine. The calcite in travertine is further divided into two mineral generations, namely calcite in primary travertine and calcite in secondary travertine. Therefore, calcite is divided into three generations from the generation of the parent to the descendants, i.e., calcite in the Mesozoic carbonate rock, calcite in the Late Cenozoic travertine, and calcite re-precipitated after travertine leaching.

The calcite in the Mesozoic limestone is micritic, microcrystal and sparry calcite, ranging in size from centimeters to micrometers. CaO and MgO in the rocks composed of these calcite are close to the theoretical value, and the other components are very low, which belong to the soluble carbonates. These calcite became the parent generation in the whole calcite evolution sequence, and they provided the material source for the calcite of later generations after being leached. The calcites of the descendants form the different morphologies of the cold-water travertine. During the formation process, physical, chemical, biological and other factors participate in the diagenesis. Among these travertines, no matter what color, except for calcite, other minerals hardly develop. The calcites of the descendants of travertine are very numerous and complex. Here, we mainly listed two of them, which are calcite in the laminal travertine and calcite in the porous travertine, because these are the main components of most travertines. The calcite of the laminal travertine is long columnar and slablike. The brown and white calcite is continuously growing without interruption. These characteristics are very different from those observed on the eye assay, which indicates that the calcite growth in the dry and cold seasons is continuous (Wang et al. 2014). On the other hand, it reflects that the hydrodynamic conditions are very stable, and the water layer is very thin with little or no biological involvement. The calcite in the porous travertine tends to be granular, and the particle size is much smaller than that of the laminal travertine, and its particle size is generally less than 100 μm. These characteristics reflect the rapid crystallization of calcite, Due to the strong hydrodynamics and the participation of biological effects, calcite cannot be continuously grown, but suddenly nucleates and grows to a certain extent then no longer grows.

The last generation of calcite is the secondary calcite in travertine. The ancestral body of this type of calcite is the deposited travertines, which are dissolved in the water by weathering and leaching, then the calcite re-precipitates through the deposition of a parent-like travertine. These calcites will adhere to the cracks, edges and even the surface of the primary calcite.

4 Conclusions

The calcite in the Huanglong cold-water travertine can be divided into three mineral generations, which reveal two different sedimentary environment systems respectively. They are the generations of the marine carbonate rock diagenesis system, and the descendant generation is the continental freshwater karst sedimentary system. Unlike conventional weathering, which converts terrestrial carbonate rocks to the ocean phase, this is done in the opposite direction. The study of different calcite mineral generation can reconstruct the paleo-environment and paleo-climate of its sedimentation.

Acknowledgements. This research was supported by National Natural Science Foundation of China (Grants nos. 41572035, 41603041 and 41877288), the Open Funds of Key laboratory of mountain hazards and surface processes (grant No. 19zd310501) and Longshan Talents program of Southwest University of Science and Technology (18lzx663).

References

Pentecost A (1995) The quaternary travertine deposits of Europe and Asia Minor. Quaternary Sci Rev 14(10):1005–1028

Wang HJ, Yan H, Liu ZH (2014) Contrasts in variations of the carbon and oxygen isotopic composition of travertines formed in pools and a ramp stream at Huanglong Ravine, China: implications for paleoclimatic interpretations. Geochimica et Cosmochimica Acta 125:34–48

Wang FD, Dong FQ, Zhao XQ (2018) The large dendritic fissures of travertine dam exposed by Jiuzhaigou earthquake, Sichuan, southwestern China. Int J Earth Sci 107(8):2785–2786

49

Cs Leaching Behavior During Alteration Process of Calcium Silicate Hydrate and Potassium Alumino Silicate Hydrate

K. Kuroda[1(✉)], K. Toda[1], Y. Kobayashi[1], T. Sato[2], and T. Otake[2]

[1] Graduate School of Engineering, Hokkaido University, Hokkaido, Japan
k7927k@eis.hokudai.ac.jp

[2] Faculty of Engineering, Hokkaido University, Hokkaido, Japan

Abstract. Zeolite, used to remove Cs from a contaminated water, would be solidified for the safety disposal. Recently, geopolymer is considered as a new binder for disposal. Geopolymer has an advantage that primary phases such as potassium almino silicate hydrate (K-A-S-H) may sorb radioactive nuclides. In this study, Cs adsorption, co-precipitation and desorption experiment were conducted, and C-S-H, which is primary phases of cement, were also employed for experiments for comparison. From these experiments, it is obtained that K-A-S-H has higher adsorption capacity of Cs than C-S-H. Cs adsorption ratio and co-precipitation ratio by C-S-H were almost same. Cs is likely sorbed by C-S-H thoroughly via ion exchanging. The desorption experiment demonstrated that most Cs was desorbed from C-S-H while 90% of Cs remained in K-A-S-H. Therefore, K-A-S-H has a higher retention capacity than that of C-S-H. Consequently, geopolymer is considered to be a better material in terms of Cs storage.

Keywords: Geopolymer · K-A-S-H · C-S-H · Radioactive waste

1 Introduction

After the accident at the Fukushima Daiichi Nuclear Power Station that occurred due to the The2011 off the Pacific coast of Tohoku Earthquake, contaminated water with radioactive nuclides such as cesium (Cs) have been continuously generated. Zeolite have been used for removing Cs from the contaminated water, and the spent zeolite are currently planned to be solidified for the safety storage and disposal. Recently, geopolymer is considered as a new binder for safety disposal of spent zeolite. Geopolymer has an advantage that primary phases such as potassium almino silicate hydrate (K-A-S-H) may have property for sorbing radioactive nuclides. However, there are few data about the adsorption behavior of Cs by K-A-S-H and the Cs leaching during their alteration.

2 Methods and Approaches

In this study, C-S-H, which is primary phases of cement, were also employed for experiments for comparison. In adsorption experiments, powder C-S-H and K-A-S-H were put into Cs-solution whose concentration is 1.0 mM at 298 K for a week. And in co-precipitation experiment, the materials to synthesize C-S-H were put. K-A-S-H could not be conducted co-precipitation experiment because water react with materials during synthesize. The solid sample after adsorption experiment were investigated in batch test and flow-through test as desorption experiment of Cs. The period of batch test is 4months and that of flow-through test is a month. And deionized water was used in both of them.

3 Results and Discussion

The adsorption ratio by K-A-S-H is 92%, while the adsorption ratio by C-S-H is 29%. The reason of this is considered that the size of sorption site is based on ionic radius of K or Ca, and that of Cs is similar to K than Ca. The adsorption ratio and co-precipitation ratio by C-S-H is almost same. Cs is likely sorbed by C-S-H thoroughly via ion exchanging, so it may be easy to sorb even after generation. In batch test as desorption experiment, the reaction between solid and water phase became equilibrium in 1month, and Cs concentration were almost stable after that. The desorption ratio from C-S-H was around 20% and from K-A-S-H was around 2%. But in flow through test, C-S-H desorb almost all of Cs in a day. It is considered that Cs sorption by C-S-H is ion exchange, so it is easy to leach by ion exchange too. On the other hand, The desorption ratio from K-A-S-H was almost 1% per day until 1month had past. Cs/Si ratio in each day was constant, and Si concentration is considered to depend on the dissolution amount of K-A-S-H. It is considered that Cs concentration also depended on that. From these results, it can be said that K-A-S-H has higher property to prevent desorption of Cs than C-S-H.

4 Conclusions

Consequently, K-A-S-H has higher retention capacity than that of C-S-H. These results show that geopolymer whose matrix is composed of K-A-S-H is considered to be better in terms of Cs storage.

Acknowledgements. This work was supported by MEXT 8桁の認可番号, Long-term performance of cement disposal systems for synthetic zeolites and titanates arising from reprocessing of contaminated water.

50

Geochemical Behavior of Heavy Metals During Treatment by Phosphoric Fertilizer at a Dumping Site in Kabwe, Zambia

H. Kamegamori[1(✉)], K. Lawrence[1], T. Sato[2], and T. Otake[2]

[1] Graduate School of Engineering, Hokkaido University, Sapporo, Japan
skj-9mm@eis.hokudai.ac.jp

[2] Faculty of Engineering, Hokkaido University, Sapporo, Japan

Abstract. Kabwe area in Zambia has been affected by heavy metal contaminations which derived from past mining activities. Particularly, Pb is one of the most concerned elements for human health in Kabwe. In this context, treatment by phosphoric fertilizer was conducted to reduce Pb solubility in soil and slag, limiting their bioavailability. Because leach plant residue in Kabwe contains metal sulfate minerals with high solubility, concentration of heavy metals in groundwater is high. We clarified the geochemical behavior of heavy metals (Pb, Cd, Zn and Cu) after the addition of phosphoric fertilizer (Triple Super Phosphate: TSP) in column experiment. Immobilization of Pb and Cd lowers concentration of the metals in ground below WHO environmental standard.

Keywords: Insolubilization · Soil amendments · Heavy metal contamination · Phosphate mineral · Mine waste

1 Introduction

Kabwe town is the worst polluted place in Africa due to mining and smelting of Pb and Zn ores. Orthophosphate has been receiving a lot of attention as stabilization agent for heavy metals, In order to reduce dispersion and mobility of Pb metal from the slag, we suggests treatment by adding phosphoric fertilizer (Triple super phosphate: TSP) which is effective and locally available.

2 Methods and Approaches

We conducted a series of column experiments in 50 ml of syringe tubes, simulating treatment for stacked slags at a dumping site in Kabwe. The syringes were filled with slags obtained from Kabwe site with 10 g of TSP on the top of slag sample. 6 mL of rain water obtained from the site was added every day, which is consistent with average daily precipitation rate. Infiltrated water was collected at the bottom of syringe and analyzed by ICP-AES and ICP-MS. After the column experiments, the slag samples in the column also investigated to understand geochemical processes occurred during the experiments by SEM/EDS.

3 Results and Discussion

We confirm the reduction in Pb and Cd concentrations in the eluents. Remarkably, the reduction for Pb concentration is 96%. In the infiltrated slags simultaneously, we observe the alteration from $PbSO_4$ to $(Pb, Ca)_5(PO_4)_3Cl$ (Fig. 1), which effects to reduce the mobility of Pb. In contrast, elution of Zn and Cu from the slags are promoted by the presence of TSP. This is due to lowering pH by TSP, desorbed Zn and Cu from amorphous and crystalline iron hydroxides. It suggests to supply orthophosphate at neutral pH range is effective for immobilization of heavy metals in slags.

Fig. 1. The alteration from anglesite to pyromorphite

4 Conclusions

We confirmed the behavior of some heavy metals applied TSP in column scale. From the results, TSP could immobilize Pb and Cd, however, it promoted elution of Zn and Cu due to soil acidification. This suggests applying TSP with dolomite to the slag could be a better remediation method.

Acknowledgements. This study is supported by International Collaborative Research Program (SATREPS): Visualization of Impact of Chronic/Latent Chemical Hazard and Geo-Ecological Remediation in Zambia. I'm deeply grateful to Dr. Kasama who taught me how to use SEM/EDS in Center for Electron Nanoscopy, Denmark Technical University.

51

Utilization of Associated Oil Gas: Geo-Ecological Problems and Modernization of the State

L. Z. Zhang[1,2] and H. Y. Sun[1,3](✉)

[1] Department of Mineral Developing and Oil & Gas Engineering, RUDN University, Moscow, Russia
657273629@qq.com
[2] Liaoning Shihua University, Fushun, China
[3] Qinhuangdao Experimental Middle School, Qinhuangdao, China

Abstract. In the world vast of oil is extracted, especially in China. Respectively produce associated petroleum gas is in a large volume. There are geo-ecological problems in the utilization of associated petroleum gas. In connection with the increasing requirements for the preservation of the state of the biosphere in China, the process of modernization was begun. Chinese modernization of associated petroleum gas utilization is presented.

Keywords: Associated petroleum gas · Modernization · Technology · Geo-ecological problems

1 Introduction

Associated petroleum gas (after this APG) is a mixture of various gaseous hydrocarbons dissolved in oil and released in the process of extraction and preparation of oil. The oil gases also include gases released in the operations of thermal processing of oil (cracking, reforming, hydrotreating, etc.), consisting of saturated and unsaturated (methane, ethylene) hydrocarbons.

From geology, APG is often formed during the Ordovician and Silurian periods. Sometimes it is built late to the Cretaceous. (Vorobiev and Zhang 2018).

2 Methods and Approaches

The PRC's world ranking in oil production is quite high and, accordingly, the volumes of simultaneously produced associated gas are very significant. In 2011–2013 in China, APG was provided in the amount of 27.3, 28.9, 30.2 billion m^3. In addition, the share of production took more than 5.1% in the world.

Previously, in China, APG was traditionally considered not as a valuable resource, but as a by-product of oil production, the simplest method of which utilization is flaring in many fields, especially in Northeast China.

3 Results and Discussion

Associated gas recovery technology using membrane separation is based on the following steps:

- removing micro solid particles, crude oil and heavy hydrocarbon emulsion contained in associated gas;
- after preliminary impurity removal, heating up to 590 °C in a heat exchanger;
- introducing into a liquid rotary compressor;
- introducing the heated gas into a desulfurization tank, and desulfurizing;
- introducing the desulfurized gas into a membrane separator, and separating;
- introducing the gas from the membrane separator to a molecular sieve tank, and performing deep desulfurization and decarburization;
- cooling to obtain the product (Mo 2013).

This technological process is quite simple and convenient in industrial operation, and besides, it is characterized by low operating costs, high recovery rate and can be, after a little adaptation, widely circulated.

4 Conclusions

Associated petroleum gas will become a valuable raw material for further processing.

China's economy needs to use APG to reduce greenhouse gas emissions.

Modernization of processing of APG in China and its prospects lie in the area of increasing the efficiency of processing of associated gas, reducing energy consumption in the course of processing, flexible operation, convenient installation and operation.

References

Mo JL (2013) Recovery process for petroleum associated gas. China Patent CN102994180A, 27 March 2013

Vorobiev AE, Zhang LZ (2018) Apply innovative technologies for processing of associated gas in China. Eurasian Sci. J. 10(2)

52

Environmental Solutions for the Disposal of Fine White Marble Waste

I. Shadrunova, T. Chekushina(✉), and A. Proshlyakov

Academic N.V. Melnikov Institute of Problems of Comprehensive Exploitation of Mineral Resources, Russian Academy of Sciences, Moscow, Russia
tvche.2016@gmail.com

Abstract. The article deals with environmental problems of formation of fine white marble wastes on the territory of Koelga deposit and total mining complex. An inventory analysis of marble waste was carried out, environmental assessment of fine marble waste and their impact on the ecology of the complex territory was carried out and theoretically justified. Planned and scientifically justified ways of large-scale utilization in the production of ceramic bricks.

Keywords: Formation of fine marble waste · Environmental assessment · Amount of waste · Waste disposal

1 Introduction

The growth of industrial and mining production, the progress of civilization increase environmental problems due to the increasing consumption of mineral and other resources from the bowels of the Earth, due to the rapid rise in the number of solid man-made wastes of different productions. These wastes can be used for the production of building materials and to improve the environmental safety of mining regions. Abandoned lands are exempted from waste dumps and territories have environmental and economic benefit (Oreshkin 2017).

2 Methods and Approaches

In the world and in the Russian Federation there are the technologies of extraction and processing of non-metallic mineral resources. During these technological processes man-made wastes are formed. Their queries and the surrounding areas are withdrawn from economic circulation, violate natural landscapes - their man-made options are created. It destroys the soil, changes modes of rivers, lakes, reservoirs, underground and surface groundwater and causes great damage to the environment (Meshheryakov et al. 2009).

Koelga deposit of white marble began the work from 1924. During this time in the dumps huge amounts of fine wastes of extraction and processing marble were accumulated (Tseytlin 2012).

Almost all kinds of new productions require new construction, materials and mineral resources. Therefore, to improve the environmental safety of the regions it is necessary to carry out comprehensive development of deposits, and also to utilize man-made wastes in the production of building materials. To solve the above problems a comprehensive environmental assessment of man-made waste requires. The assessment should include amount of accumulated volumes for large-scale utilization of man-made waste in the production of building materials, products (Khokhryakov et al. 2013).

The purpose of the article is the ecological assessment of formation of man-made waste products of mining production in the form of white marble with a decrease in the available subsoil mineral resources for the production of building materials and products.

To achieve the goal, it is necessary to justify the use of these wastes as raw materials for the production of building materials and products. This will simultaneously improve the environmental safety of the territories due to large-scale utilization of man-made waste marble and will free up the areas occupied by dumps.

3 Results and Discussion

It was calculated that in 2018 the total mass in the dumps is more than 25 million tons of fine marble wastes, and the area of dumps - more than 20 hectares. An important task was also the calculation of the environmental damage from the abandoned territories under the dumps of fine marble waste.

To calculate it was analyzed the environmental effect from their utilization by reducing the area under the dumps, and the pollution of the environment of mining complex territory (Fadeichev et al. 2012).

When calculating it was determined that for the Chelyabinsk region the damage to the environment from storage of fine marble waste in dumps is about 500 thousand rubles a year (in the prices of 2018). Taking into account the amount of wastes already placed in dumps of JSC "Koelgamramor", the environmental damage will amount to over 30 million rubles.

According to calculations the utilization of fine marble waste in brick production will significantly reduce the environmental load on the environment. It will take place by reducing waste mass in dumps, that will allow to reduce the abandoned areas under dumps and to return the land to use.

In the articles it was determined the amount of recyclable fine marble waste at 1 m^3 of molding mixture for the production of ceramic bricks of multiple colors: terracotta or dark brown; light red or pinkish; fawn or straw.

The analysis of the results of technical tests showed that, on the basis of fine marble wastes, it is possible to obtain ceramic bricks of danger class 4, which corresponds to state standard GOST of the Russian Federation. It was found that burning of over-moulded ceramic raw makes at temperature 850… 900 °C. It is proved that at that temperature, the particles of marble are not affected by the process of decarbonization. Therefore, there is no greenhouse gas emissions - carbon dioxide, i.e. ecology of this mining territory is not the subject to harmful effects. Moreover, as above stated, the level of danger of marble wastes was higher (class 3) by one step, than the level of danger of the ceramic bricks produced (4 class) on the basis of these wastes. Also amounts of energy for manufacture of these ceramic bricks reduced significantly as compared to common ceramic brick (Moumouni et al. 2016).

4 Conclusions

Thus, the total mass accumulated fine marble wastes and environmental damage to mining area of Koelgo deposit were determined. The technology of improving the ecology of the region due to large-scale utilization of the above marble wastes in the production of ceramic bricks was elaborated. The possible number of bricks of different colors at full disposal of accumulated marble wastes was determined. Using environmental life cycle assessment of finished products based on fine marble wastes the possibility of obtaining an environmentally safe effective bricks was theoretically justified and the technology of their production was elaborated. The dependence of color products from fine marble waste was defined. So, at an amount of 20% of fine marble waste in the mixture by mass of clay rocks the ceramic brick has a dark brown color, and at 40% - has straw color. The influence of the elemental composition of the mixture on the color of the brick was determined (Merem et al. 2017).

It was proved that the most environmentally safe, resource-saving way of man-made waste disposal is their utilization in the production of building materials and products. This method releases territories abandoned for storing waste and provides environmental and economic effects from the elimination of dumps (Bilgin et al. 2012).

Thus, the environmental problems of the Russian Federation connected with rise of man-made waste of white marble with a decrease in available reserves of mineral resources for the production of building materials and products were specified. The scientific foundations of the integrated environmental assessment methodology of man-made wastes and their large-scale utilization in the production of building materials and products were elaborated (Hebhoub et al. 2011). The possibility to use these wastes as raw components for their production, while solving environmental problems of the territories due to large-scale utilization of man-made waste was justified. This extends the raw material base and contributes to the integrated development of bowels, their mineral and man-made resources. Utilization of man-made waste allows to get a huge environmental and economic effects on the territory of the Russian Federation.

References

Bilgin N, Yeprem HA, Arslan S, Bilgin A, Günay E, Mars MO (2012) Use of waste marble powder in brick industry. Constr Build Mater 29:449–457

Fadeichev AF, Khokhryakov AV, Grevcev NV, Cejtlin EM (2012) Dynamics of negative impact on the environment at different stages of mining development. News High Educ Inst Mountain Mag 1:39–46

Hebhoub H, Aoun H, Belachia M, Houari H Ghorbel E (2011) Use of waste marble aggregates in concrete. Constr Build Mater 25(3):1167–1171

Khokhryakov AV, Fadeichev AF, Cejtlin EM (2013) Application of an integral criterion for determining the environmental hazard of mining enterprises. News Ural State Mining Univ 1:25–31

Merem EC et al (2017) Assessing the ecological effects of mining in West Africa: the case of Nigeria. Int J Mining Eng Mineral Process 6(1):1–19

Meshheryakov YuG, Kolev NA, Fedorov CV, Suchkov VP (2009) Stroymaterialy Production of granulated phosphogypsum for the cement industry and building products, vol 5, pp 104–106

Moumouni A, Goki NG, Chaanda MS (2016) Natural Resources Geological exploration of marble deposits in Toto Area, Nasarawa State, Nigeria, vol 7, pp 83–92

Oreshkin DV (2017) StroymaterialyInvironmental problems of integrated development of mineral resources in the large-scale utilization of man-made mineral resources and waste in the production of building materials, vol 8, pp 55–63

Tseytlin EM (2012) Features of environmental hazard assessment of mining enterprises Theses of the report of VII Krakow conference of young scientists. AGH University of Science and Technology, Krakow, pp 809–819

53

Murataite-Pyrochlore Ceramics as Complex Matrices for Radioactive Waste Immobilization: Structural and Microstructural Mechanisms of Crystallization

S. Krivovichev[1,2(✉)], S. Yudintsev[3], A. Pakhomova[4], and S. Stefanovsky[5]

[1] Kola Science Center, Russian Academy of Sciences, Apatity, Russia
krivovichev@admksc.apatity.ru
[2] Department of Crystallography, St. Petersburg State University, St. Petersburg, Russia
[3] Institute of Geology of Ore Deposits, Petrography, Mineralogy, and Geochemistry, Russian Academy of Sciences, Moscow, Russia
[4] Deutsches Elektronen-Synchrotron (DESY), Petra III, Hamburg, Germany
[5] Frumkin Institute of Physical Chemistry and Electrochemistry, Russian Academy of Sciences, Moscow, Russia

Abstract. Murataite-pyrochlore titanate ceramics are attractive waste forms capable to immobilize radioactive waste streams of complex compositions, thus eliminating the need for further chemical separation. We have investigated structures of three types of murataite: 3*C*, -5*C*, and -8*C* phases and demonstrate their polysomatic nature and structural complexity. Structurally simple pyrochlore crystallizes first, followed by crystallization of murataite-5*C* containing pyrochlore cells surrounded by fragments of Keggin clusters. This phase is overgrown by murataite-8*C* containing both murataite and pyrochlore cells. The crystallization finishes with the formation of murataite-3*C*, which is the most stable and less actinide-rich. The microstructure formed via this mechanism reminds a Russian doll, which creates additional barrier for the actinide leaching from the pyrochlore core. The high chemical and structural complexity of the pyrochlore-murataite series is unparalleled in the world of crystalline materials proposed for the HLRW immobilization, which makes it unique and promising for further exploration.

Keywords: Murataite · Pyrochlore · Crystal structure · Crystallization · Microstructure · Radioactive waste · Actinides

1 Introduction

One of the most important tasks for the advanced nuclear cycle is the elaboration of waste forms capable to immobilize waste streams of complex compositions, thus eliminating the need for further chemical separation. In this regard, the murataite-pyrochlore titanate ceramics attract considerable attention due to their ability to immobilize radioactive wastes with different and complex chemical compositions, including actinides such as Pu-238. Over last few years, there has been a renewed interest in their synthesis and investigations (Maki et al. 2017; Lizin et al. 2018, etc.).

Murataite-(Y) is a complex titanate mineral first discovered in alkali pegmatites in St. Peters Dome area in Colorado, United States and later found in pegmatites in the Baikal region in Russia. Its crystal structure (cubic, space group *F*-43 *m*, $a = 14.886$ Å) was determined by Ercit and Hawthorne (1995) as based upon a framework of corner-linked α-Keggin clusters hosting a complex metal-oxide substructure. The simplified formula of natural murataite-(Y) can be written as ${}^{[8]}R_6^{[6]}M1_{12}^{[5]}M2_4^{[4]}TX_{43}$, where R = Y, HREE, Na, Ca, Mn, *M1* = Ti, Nb, Na, *M2* = Zn, Fe, Ti, Na, T = Zn, Si and X = O, F, OH. The interest in murataite-(Y) was renewed in 1982, when its synthetic analogue was identified in Synroc-type titanate ceramics with imitators of high-level radioactive waste at the Savannah River nuclear power plant (Morgan and Ryerson 1982). Laverov et al. (1998) reported the formation of murataite-type titanate phase in the uranium-bearing Synroc matrix from the Mayak factory, a radiochemical facility for the reprocessing of nuclear fuel located in Southern Ural, Russian Federation. It was found that five volume percent of synthetic murataite accumulate about 40% of the total amount of uranium present in the sample, which led to follow-up detailed studies of chemistry and properties of this material. Transmission electron studies allowed identification of synthetic varieties of murataite with $3 \times 3 \times 3$, $5 \times 5 \times 5$, $7 \times 7 \times 7$ and $8 \times 8 \times 8$ fluorite-like cubic supercells, referred in the following as murataite-3*C*, -5*C*, -7*C* and -8*C* phases (Laverov et al. 2011).

2 Methods and Approaches

We have studied crystal structures of murataite-3*C*, -5*C* and -8*C* using single-crystal X-ray diffraction analysis on the samples obtained by melting the mixture of oxides in an electric furnace at 1500 °C with subsequent cooling to the room temperature as described by Laverov et al. (1998). The details of the experimental procedures used to obtain structure models have been described in detail in (Krivovichev et al. 2010; Pakhomova et al. 2013, 2016).

3 Results and Discussion

Urusov et al. (2005) proposed that synthetic murataites can be considered as members of murataite-pyrochlore polysomatic series consisting of different combinations of 2D modules. The structural determination of murataite-5*C* reported by Krivovichev et al. (2010) confirmed the assumption about the modular nature of the polysomatic series and demonstrated that the murataite- and pyrochlore-type modules are not layers but zero-dimensional blocks (nanoscale clusters), combination of which in a 3-dimensional space generates at least two different derivative structures, which combine structural features of both murataite and pyrochlore. In particular, the crystal structure of murataite-5*C* can be described as an ordered arrangement of pyrochlore unit cells immersed into the recombined murataite matrix, i.e. a substructure consisting of murataite structure elements.

The crystal structure of murataite-3*C* was reported by Pakhomova *et al.* (2013), who demonstrated its general identity to natural murataite, with some important chemical and structural modifications. The modular nature of the murataite-pyrochlore polysomatic series was discussed by Laverov et al. (2011).

The crystal structure of murataite-8*C* was reported by Pakhomova et al. (2016) as based upon a three-dimensional octahedral framework that can be described as an alternation of murataite and pyrochlore modules immersed into transitional substructure that combine elements of the crystal structures of murataite-3*C* and pyrochlore. The obtained structural model confirmed the polysomatic nature of the pyrochlore-murataite series and illuminated the chemical and structural peculiarities of crystallization of the murataite-type titanate ceramic matrices.

Table 1. Information-based structural complexity parameters for the synthetic members of the pyrochlore-murataite polysomatic series

Material	v [atoms]	I_G [bits/at.]	$I_{G,total}$ [bits/cell]
Pyrochlore	22	1.686	37.088
Murataite-5C	336	4.892	1643.840
Murataite-8C	1387	6.558	9096.031
Murataite-3C	71	3.226	229.044

4 Conclusions

The information-based complexity parameters for the members of the pyrochlore-murataite are listed in Table 1. Both kinds of information-based complexity parameters (per atom and per cell) behave in a similar fashion: they are relatively small for the initial pyrochlore phase, then increase for murataite-5*C*, reach their maxima for murataite-8*C* and decrease for the final murataite-3*C* phase. This trend is also followed in the crystallization of murataite-pyrochlore ceramics: structurally simple and actinide-rich pyrochlore crystallizes first, creating conditions for the saturation of melt with Keggin clusters, which triggers crystallization of murataite-5*C* containing pyrochlore unit cells surrounded by fragments of Keggin clusters. This pyrochlore-rich phase is overgrown by murataite-8*C* containing both murataite and pyrochlore unit cells. The crystallization finishes with the formation of pure Keggin phase murataite-3*C*, which is the most stable and less actinide-rich. The microstructure formed during such a crystallization reminds a Russian doll ('matryoshka'), which creates additional barrier for the actinide leaching from the pyrochlore (or crichtonite) core. The high chemical and structural complexity of the members of the pyrochlore-murataite series is unparalleled in the world of crystalline materials proposed for the high-level radioactive waste immobilization, which makes it unique and promising for further technological and scientific exploration.

Acknowledgements. This work was supported by the President of the Russian Federation grant for leading scientific schools (grant NSh-3079.2018.5 to SVK).

References

Ercit TS, Hawthorne FC (1995) Murataite, a UB_{12} derivative structure with condensed Keggin molecules. Can Mineral 33:1223–1229

Krivovichev SV, Yudintsev SV, Stefanovsky SV, Organova NI, Karimova OV, Urusov VS (2010) Murataite-pyrochlore series: a family of complex oxides with nanoscale pyrochlore clusters. Angew Chem Int Ed 49:9982–9984

Laverov NP, Sobolev IA, Stefanovskii SV, Yudintsev SV, Omel'yanenko BI, Nikonov BS (1998) Synthetic murataite: a new mineral for actinide immobilization. Dokl Earth Sci 363:1104–1106

Laverov NP, Urusov VS, Krivovichev SV, Pakhomova AS, Stefanovsky SV, Yudintsev SV (2011) Modular nature of the polysomatic pyrochlore-murataite series. Geol Ore Dep 53:273–294

Lizin AA, Tomilin SV, Poglyad SS, Pryzhevskaya EA, Yudintsev SV, Stefanovsky SV (2018) Murataite: a matrix for immobilizing waste generated in radiochemical reprocessing of spent nuclear fuel. J Radioanal Nucl Chem 318:2363–2372

Maki RSS, Morgan PED, Suzuki Y (2017) Synthesis and characterization of a simpler Mn-free, Fe-rich M3-type murataite. J Alloys Compd 698:99–102

Morgan PED, Ryerson FJ (1982) A "cubic" crystal compound. J Mater Sci Lett 1:351–352

Pakhomova AS, Krivovichev SV, Yudintsev SV, Stefanovsky SV (2013) Synthetic murataite-3C, a complex form for long-term immobilization of nuclear waste: Crystal structure and its comparison with natural analogues. Z Kristallogr 228:151–156

Pakhomova AS, Krivovichev SV, Yudintsev SV, Stefanovsky SV (2016) Polysomatism and structural complexity: Structure model for Murataite-8C, a complex crystalline matrix for the immobilization of high-level radioactive waste. Eur J Mineral 28:205–214

Urusov VS, Organova NI, Karimova OV, Yudintsev SV, Stefanovskii SV (2005) Synthetic "murataites" as modular members of a pyrochlore-murataite polysomatic series. Dokl Earth Sci 401:319–325

Security Test of New Technology in view of Increased Performance of Oil Platforms without Increasing Environmental Risks

E. M. Tanoh Boguy(✉) and T. Chekushina

Department of Mineral Developing and Oil & Gas Engineering, Engineering Academy, RUDN University, Moscow, Russia
boguymartialeddy@gmail.com

Abstract. In this article, it will be important to note the context of the gradual depletion of existing fields, which are pushing to expand research and exploitation of new fossil fuel resources in order to meet the growing demand for fuel and, despite international regulations to combat global warming. Consequently, an increase in offshore platforms in global hydrocarbon production to compensate for the depletion of the earth's reserves is becoming a major problem for the oil industry. Given the financial unforeseenness that is represented, and the energy autonomy which is provided, marine exploitation has become a problem for states with a large sea area and, therefore, an environmental.

Keywords: Security test · New technology · Increased performance · Environmental risks

1 Introduction

The use of new reserves, in economically viable conditions depends on the available technologies. The development of deep and ultra-deep offshores requires considerable research and development efforts. Progress has also been made in managing the multiple risks associated with this activity. A disaster like «Deepwater Horizon» led to a detailed analysis and sharing of findings by industry experts.

Despite the security rules on the platforms, in fact, some major incidents are revealed, the causes of which are multifactorial in nature and which have dire consequences for both humans and the environment.

Legislation forces organizations to take responsibility for dealing with disasters, which has been developed over time and in different ways in different countries.

Our analysis of how environmental risks are taken into account by various subjects and offers development prospects to ensure better safety of offshore activities.

2 Methods and Approaches

Some of the accidents at oil rigs, such as the Deepwater Horizon, in the spring of 2010 caused a shock wave in their magnitude and severity that convinced that such accidents could occur. In fact, some states have taken steps to raise the level of security. In fact, some states have taken upon themselves the task of "solving the problem of providing security on the shelf." Global hydrocarbon production is becoming more and more offshore, accounting for more than 35% of oil and 19% of gas. Since the deposits are located at great depths, states and companies must develop the potential for their use, seeking to control the risks inherent in this activity carried out under extreme conditions. In order not to have restrictions in the conditions in which exploration, drilling and mining operations are carried out, by more and more complex and risky methods. In addition, oil companies are well aware of their interest in investing in the development of new technologies. This allows them to gain an industrial competitive advantage in strategic areas of deepwater exploitation. Significant progress has been made in managing the multiple risks inherent in offshore operations. Oil companies put prevention at the level of operating conditions.

3 Results and Discussion

Regardless of the achievements observed, it is obvious that the safety rules applied on the platforms guarantee greater efficiency than environmental protection, and that more and more risks are encountered. Increasing risks to humans and the environment is inextricably linked with the complexity of drilling operations. Working platforms continue to find solutions that completely avoid any potential risks in the protected areas. Advanced technology and security measures suggest that there is a clear improvement, but the limit between politically correct and pollution is quickly exceeded when it comes to such profits.

4 Conclusions

Finally, it is important to include risk management in determining policies, procedures and plans, as well as specific risk mitigation measures that will be taken to manage security risks. The environment is associated with all sorts of accidents, while drilling and operating the platform. The accident, which is a major problem for these exploitation will be a hydrocarbon spill, which is highly unlikely and will be limited to pumping oil and fuel stored on support vessels in the event of a tank failure or reloading pipe.

Cultural Heritage, Artifacts and their Preservation

55

Monitoring of the State of St. Petersburg Stone Monuments and the Strategy of their Preservation

O. Frank-Kamenetskaya[1(✉)], D. Vlasov[1], V. Rytikova[2], V. Parfenov[3], V. Manurtdinova[2], and M. Zelenskaya[1]

[1] St. Petersburg State University, St. Petersburg, Russia
ofrank-kam@mail.ru
[2] State Museum of Urban Sculpture, St. Petersburg, Russia
[3] St. Petersburg Electrotechnical University «LETI», St. Petersburg, Russia

Abstract. The results of the multi-year monitoring of the state of Saint Petersburg stone monuments are summarized. The unique collection of decorative stones in museum Necropolis and the deposits that were most likely used to create them are studied. The processes of stone monuments' degradation in response to physical, chemical and biogenic influences are discussed. Special attention is paid to describing the monitoring methodology and the structure of the monitoring information database. Drawing on received results, the strategy for the conservation and restoration of monuments are discussed. The obtained data are of exceptional scientific interest in studying the processes of stone deterioration under the impact of the environment.

Keywords: Cultural heritage · Monitoring · Stone deterioration · Anthropogenic weathering · Restoration and conservation works

1 Introduction

Preservation of the monuments of cultural heritage is one of the priorities of the modern society. This problem becomes especially acute where the monuments are exhibited in the open air and subjected to destructive effects of the environment. In large cities, such as St. Petersburg, the deterioration of natural stone is notably fast, which is primarily due to the influence of the anthropogenic factor (The Effect 2019). Now we present the results of a multi-year, comprehensive study of the state of historical stone monuments of St. Petersburg, which are exposed to the destructive impact of the urban environment. The obtained data are of exceptional scientific interest for studying the processes of stone deterioration under the impact of the environment.

2 Methods and Approaches

Monitoring studies have been carried out in the Historical Necropoleis of the Museum of Urban Sculpture since 1998, where on a small square there is a unique collection of decorative and facing stone. The stone is intensively destroyed due to destructive influence of the volatile and humid Petersburg climate and unfavorable ecological situation. In this work, in addition to the Saint Petersburg scientists, museum staff and restorers, post-graduate students and students of the St. Petersburg State University, the Russian and Herzen State Pedagogical University took part. Over the past years, more than 1300 monuments of the Museum Necropolis have been examined (some of them several times), Based on the results obtained, a methodology for monitoring studies of stone materials of monuments was developed, which included the following steps: 1. Visual inspection of the object. Photographic documentation. Sampling. 2. Qualimetric evaluation of the integral state of the monument material (performed for 348 monuments). 3. Mapping of the types of material deterioration. 4. Examination of the samples of material and products of its deterioration by instrumental procedures (petrographic description of thin sections under a polarizing microscope, SE microscopy with EDX, X-ray phase analysis, biological methods). 5. Examination of the species composition of the microbial community on the surface of the monument. 6. Developing a 3D model of the monument and a quantitative estimate of the types of destruction of its material by the results of laser scanning. 7. Study of the local corrosivity of the air environment near the monuments. 8. Archival research. 9. Creating and maintaining a database on the state of the sculptural monuments in St. Petersburg.

3 Results and Discussion

Stone Material of Monuments. The diverse stone material in the museum Necropoleis is represented by marbles, limestones, granites and other hard rocks (gneisses, gabbroids, amphibolites, quartzites). The museum Necropoleis are not inferior to the historical center of St. Petersburg in the variety of stone. Basically, the stone came from Italy and the areas close to St. Petersburg (from the territory of the present Leningrad region, Karelia and Finland).

Qualimetric Evaluation of the Integral State of the Monument Material. The technique was developed jointly with V.M. Marugin (VITU, SPb). It was shown that the degree of stone destruction in the museum Necropoleis varies from 2 to 51%. In most cases, the extent of carbonate rock deterioration does not exceed 25% and that of granite and other hard silicate rocks - 10%. This is due to the considerable contribution of chemical weathering (formation of gypsum-enriched patina) in the deterioration of memorials of marble and limestone. Cracks occur on the surface of carbonate rocks that are heterogeneous in composition and structure (Ruskeala, Italian breccia and brecciated marbles, Pudost and Putilovo limestones) at least 10% more often than on other denser and more homogeneous marbles and limestones. But on denser solid silicate rocks (granites, etc.), cracks occur no less frequently than on carbonate rocks. At the same time, they are much more common (found on 80% of monuments) on such dense homogeneous rocks as Serdobol granite and Shokshinsky quartzite, which indicates a possibility of their anthropogenic or constructional origin. The incidence of the primary gypsum crust on the surface of limestones is more frequent than on the surface of marbles. Among limestones, the gypsum-rich patina is most often found on the surface of the porous Pudost travertine (on the surfaces of 50% of examined monuments). Among marbles, it is most often seen on the homogeneous Carrara marble (on the surfaces of 26% of the surveyed monuments). Its detachment together with marble and the formation of a secondary gypsum crust are observed only on the monuments with a complex surface relief made of dense homogeneous marble: (white Carrara and light gray Bardiglio). In fouling, biofilms with dominant fungi are widespread on the surface of all rocks The input of microorganisms (fungi, algae, lichens) in rock deterioration varies from 2 to 10%. The degree of manifestation of various types of stone destruction significantly varies depending on the exhibiting conditions of the monument, the characteristics of the stone material, as well as the timing and effectiveness of work on the care.

Mapping of Deterioration Forms. Ultrasonic sounding was used to detect heterogeneities of the rock material invisible from the surface. Method for monitoring the biofouling of cultural heritage sites using computer technology allowing to register the areas of the most threatened biodeterioration sites was developed (Fig. 1). Beside 3D laser scanning method was used for to create 3D computer models and to carry out the quantitative measurements of various kinds of damage of the monument materials: cracks, chips, scratches, gypsum crusts (Fig. 2) and others.

Database on the State of the Sculptural Monuments of St. Petersburg. One of the most important stages in monitoring the state of the monuments was creating and populating a specialized database used to store, analyze and structure the accumulated factual information Currently, the database includes characteristics of the state of 650 stone monuments in the Necropoleis of the Museum of Urban Sculpture and in other parts of St. Petersburg.

Approaches and Methods of Monument Protection from Damage. To assess the effectiveness and safety of different approaches when removing biofilms, mud buildups and gypsum crusts from the surface of stone monuments, a comparative analysis was made of the potential of various chemical biocidal treatments and of laser cleaning options. The results of the experiments showed that the laser cleaning technology for removal of biofilms from the surface of the stone is comparable, and in some cases even superior to chemical treatment with hydrogen peroxide and kaolin. In the case of intensive development of biofouling, containing mosses and lichens, the efficiency of laser cleaning is significantly higher than the efficiency of chemical biocidal treatment. The use of laser cleaning to remove gypsum-rich patina is also effective.

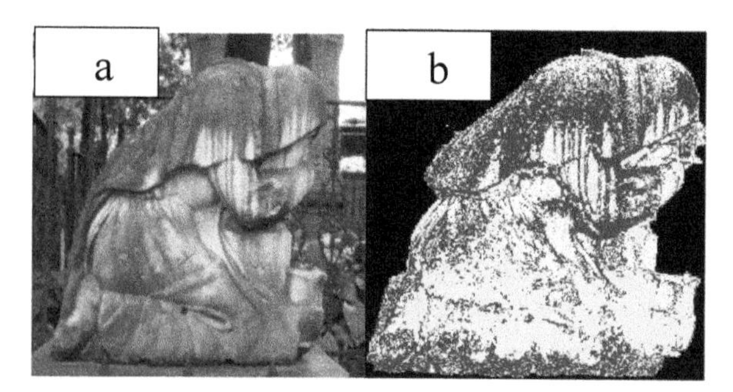

Fig. 1. Biofilms with dominant algae (color yellow) and dark-colored micromycetes (color brown): a-photograph, b-map

Fig. 2. Electronic 3D-model of fragment of the mourner sculpture, on which the area of the gypsum crust is highlighted

4 Conclusion

Integrated monitoring the state of St. Petersburg monuments provided an objective picture of the state of the their materials, makes it possible to take timely interventions for the restoration and conservation of works of art, to plan the necessary measures to protect the stone from deterioration and in result make it possible to preserve and adequately exhibit the works of monumental sculpture and memorial art of St. Petersburg, which are an impressive, imaginative part of the world history and culture.

Acknowledgements. This study was supported by RSF project no 19-17-00141 and performed using the equipment of the SPBU resource centers "X-Ray Diffraction Methods for Studying Matter," "Nanotechnologies," and "Geomodel".

Reference

Frank-Kamenetskaya OV, Vlasov DY, Rytikova VV (Ed) (2019) The Effect of the Environment on Saint Petersburg's Cultural Heritage. Springer, Switzerland

56

Identifying the Decorative Stone Samples from the Mining Museum's Collection: First Results

N. Borovkova[1(✉)] and M. Machevariani[2]

[1] Mining Museum, St. Petersburg Mining University, St. Petersburg, Russia
borovkova_nv@pers.spmi.ru
[2] Assistant of the Department of Mineralogy, Crystallography and Petrography, St. Petersburg Mining University, St. Petersburg, Russia

Abstract. The report presents the primary results of a study of a unique collection of polished decorative stone samples belonging to Empress Catherine the Great. Primary macroscopic analysis of 83 plates, divided into 13 groups according to similar features, was performed. The bulk chemical composition of rocks was estimated on the basis of XRF- analysis data, performed using a Delta Olympus XRF portable analyzer. Preliminary studies allowed to outline the characteristic fields of the studied samples of decorative rocks on the ternary plots of their bulk composition. In the future, it is planned to perform Raman spectral imaging to generate detailed maps of the mineralogical composition of the decorative stone samples.

Keywords: Polished decorative stones · Collections of Empress Catherine the great · Handheld XRF analyzer · Mining museum

1 Introduction

Natural stone serves as a unique raw material for the objects of decorative and applied arts, as well as architectural monuments. Museums and monuments of St. Petersburg store rare objects made of natural stone. The problems of their preservation are increasingly forcing restorers and art researchers to turn to geologists to identify various types of gemstone materials needed for restoration. This raises a number of problems, primarily related to the lack of reliable information about the origin of various types of such materials, as well as their accurate identification in art objects. Their study is complicated by the need to use exclusively non-destructive analytical techniques, which greatly complicates the task. Unlike European museums, in Russia, there is no complex reference collection of natural decorative (ornamental) stone, supported by current results of laboratory research, reliable information on the location, and a complete catalog of art object made of such materials. Obviously, the need for such data is highly in demand not only among restorers and art historians but also among geologists, whose research interests include the preservation of the diversity of gemstone raw materials and objects of cultural heritage.

The object of this study was one of the oldest collections of polished flat samples of natural decorative (ornamental) stone of the XVIII century, previously owned by Empress Catherine the Great (Borovkova 2017). The research contributes to the development of a methodology for evaluating historical gemstone materials of considerable cultural, museum and scientific value. In 1816, a collection was transferred from the Imperial Hermitage to the Museum of the Mining Cadet Corps (now the Mining Museum), which included collections of marbles and «rock sampled as polished plates». In the inventory of this collection, which is stored in the archives of the State Hermitage Museum, there is the following note: «Jasper and solid rocks found in the Ural Mountains, starting from Tura River by noon over the rivers Uyu and Ural» [State Hermitage archives. F. 1. O. 6l. D. 1 a-c]. The total number of such samples is not indicated, but in the catalog (1798) they are recorded from №1525 to №1725; and in the next catalog (1811), there are a number of additional samples. Thus, with the same description, samples from № 2052 to 2330 are recorded in a later catalog.

2 Methods and Approaches

Currently, more than 100 items of such decorative rocks are found in the Mining Museum. The primary macroscopic evaluation of the samples suggested wider geography of their origin. The need for reliable authentication of historical samples necessitated their thorough review.

As the first phase of the study, 83 plates were selected and broadly classified into 13 groups. The groups were formed on the basis of visual estimation of similar characteristics: the rocks structure, texture, and color. Eight of the thirteen selected groups were pre-diagnosed as known geo-referenced decorative rock types that received code names corresponding to their regional and historical affiliation, namely: Korgon porphyry; Tigiretsky breccia; Tigiretsky quartz; Kalkan, Kushkuldinskaya, Nikolaevskaya, Urazovskaya, and Surguchnaya jasper. The remaining groups require further diagnostics and are pre-defined as andesite, hornfels, marble, and green marble.

The planned research method involves the use of known geo-referenced samples. Comparative analysis of the studied samples and reference rocks will be divided into three steps: visual comparison, comparison of chemical and mineral compositions.

As mentioned above, the study of museum objects requires a special approach and the use of non-destructive techniques. In this regard, the samples chemical composition analysis was performed using a Delta Olympus XRF handheld analyzer. The measurements were carried out in the Mining mode with preliminary calibration. There was threefold spectra collection from each point with a 30 s acquisition time. In general, 50 bulk chemical composition analysis of rocks belonging to seven previously selected groups (Korgon porphyry, Tigiretsky quartz, Urazovskaya jasper, green marble, marble, andesite, and hornfels) were made.

3 Results and Discussion

The absence of reference samples let us consider the obtained results as preliminary. Nevertheless, on the ternary plots of the rock samples bulk composition, it is possible to outline the characteristic fields corresponding to certain samples groups.

The Ca-Fe-Si diagram illustrates obvious trends: visible outlining of marble, Urazovskaya jasper, and andesite characteristic fields. The bimodal distribution of points corresponding to green marble possibly occurred due to the presence of large calcite phenocrysts, the content of which varies not only during the transition from sample to sample, but also unevenly distributed over the analyzed area of the sample. On the Ca-Fe-K triangular plot, the characteristic field of Urazovskaya jasper shows that it is ferruginous which is also expressed in its characteristic purple-red color (Fig. 1). The field corresponding to the decorative andesites and the bimodality of the green marble analysis distribution are presented on all charts and can later serve as a diagnostic feature.

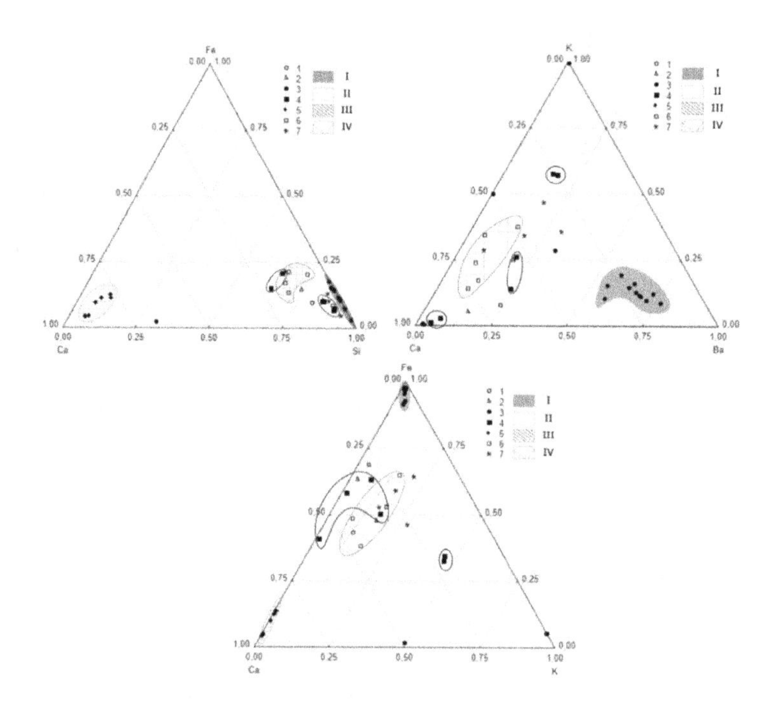

Fig. 1. Ternary plots of the bulk chemical composition of polished decorative stones from the collection of Empress Catherine the Great, plotted according to the data acquired by handheld XRF analyzer Delta Olympus. Rock types and characteristic fields: 1 - Korogon porphyry; 2 - Tigiretsky quartz; 3, I - Urazovskaya jasper; 4, II - green marble; 5, III - marble; 6 - andesite; 7 - hornfels

4 Conclusions

Thus, preliminary studies allowed us to outline the characteristic fields of the studied decorative rocks samples on their bulk composition plots. To obtain accurate results, it is necessary to expand the number of samples under study and add reference samples to the diagram. The lack of variation in the petrogenic elements content level in decorative rocks of different types and geographic reference also requires the analysis of not only chemical but also mineralogical composition of rocks. Driven by the need in non-destructive techniques, it is planned to use Raman spectroscopy to make mineralogical composition maps of the samples.

Reference

Borovkova NV (2017) Personal mineralogical collection of Empress Catherine the great in the mining museum collection. Bulletin of St. Petersburg State University of Technology and Design, № 1, Series 2. Arts Philology 8–15

57

Ceramics Sugar Jars Pieces from Aveiro Production

S. Moutinho[1(✉)], C. Costa[1,2], Â. Cerqueira[2], C. Sequeira[2], D. Terroso[2], J. Nobre[1], P. Morgado[2], A. Velosa[1,2], and F. Rocha[1,2]

[1] RISCO, Civil Engineering Department, University of Aveiro, 3810-193 Aveiro, Portugal
sara.moutinho@ua.pt

[2] GeoBioTec, Geosciences Department, University of Aveiro, 3810-193 Aveiro, Portugal

Abstract. Ceramics sugar jars pieces are, from morphological point of view, conical containers of fired clay with a hole in the vertex, which were used to sugar cane pulp maturation into sugar cake. These ceramic materials were produced in Aveiro given the existence of local raw material. Also, this occurrence of geological deposits exploited for red clays allowed the local development of strong pottery production center, transforming the city of Aveiro into one of the major Portuguese cultural heritage sites very rich in traditional ceramic tiles (*azulejos*) and other ceramic products. After local manufacture, the ceramic sugar jars pieces were exported as sugar production devices for Madeira island, Cape Verde archipelago and later, for Brazil. Also, these materials were found in buildings construction. So, this work focuses on the characterization of ceramic sugar jars produced in Aveiro and its construction use comparing with properties of other ceramics, justifying their preference for export to several countries of the world.

Keywords: Ceramic products · Sugar jars pieces · Properties · Construction elements

1 Introduction

The use of ceramic sugar jars pieces in ancient masonry walls in the Aveiro district reflects the use of these materials in construction beyond the production and transportation of sugar (Nobre 2017). Ceramic sugar jars materials were produced in Aveiro given the existence of raw material in abundance and of very good quality. In the fiftieth century the production center of Barreiro (close to Lisbon) would have ceased production and until the independence of Brazil, in the early nineteenth century, Aveiro would have been the only producer of these ceramic materials in Portugal (Morgado 2014). This production will have provided intense trade with the major sugar producing centers. Due to the local absence of natural stone for construction, the rejected/surplus ceramics were used as building material on the walls. Recently, following old house demolitions in the city of Aveiro, whole walls have been discovered with these ceramic materials, many of which were practically intact, which allowed the development of this comparative study.

2 Methods and Approaches

Mineralogical analysis was carried out by X-Ray diffraction, using a Panalytical X'Pert-Pro MPD, Kα Cu (λ = 1,5405 Å) radiation on random-oriented powders; chemical composition was assessed by X-Ray Fluorescence using a Panalytical Axios PW4400/40 X-Ray Fluorescence spectrometer for major and trace elements and Lost on Ignition (LOI) was also determined. Compressive strength was assessed by a Shimadzu: AG-IC equipment. TGA analysis was also performed.

3 Results and Conclusions

The chemical and mineralogical properties of ceramics were similar, pointing to local production using only local raw materials. Quartz is present in all samples. The phyllosilicates are not present in any sample of the sugar ceramics but are present in all the remaining ceramic samples. The presence/absence of phyllosilicates is an indicator of the heating process temperature, higher on the case of the sugar ceramic jars. The compressive strength analysis of the ceramics sugar jars pieces shows higher values (mean 9.5 MPa) than other ceramics (mean 8.0 MPa).

References

Morgado P, Rocha F (2014) Produção da cerâmica do açúcar em Aveiro pode explicar origem dos Ovos moles. Univ Aveiro J

Nobre J, Faria P, Velosa AL (2017) Paredes pão-de-açúcar em edifícios de Aveiro Evolução, materiais e características. Master thesis, New University of Lisbon

Permissions

All chapters in this book were first published by Springer; hereby published with permission under the Creative Commons Attribution License or equivalent. Every chapter published in this book has been scrutinized by our experts. Their significance has been extensively debated. The topics covered herein carry significant findings which will fuel the growth of the discipline. They may even be implemented as practical applications or may be referred to as a beginning point for another development.

The contributors of this book come from diverse backgrounds, making this book a truly international effort. This book will bring forth new frontiers with its revolutionizing research information and detailed analysis of the nascent developments around the world.

We would like to thank all the contributing authors for lending their expertise to make the book truly unique. They have played a crucial role in the development of this book. Without their invaluable contributions this book wouldn't have been possible. They have made vital efforts to compile up to date information on the varied aspects of this subject to make this book a valuable addition to the collection of many professionals and students.

This book was conceptualized with the vision of imparting up-to-date information and advanced data in this field. To ensure the same, a matchless editorial board was set up. Every individual on the board went through rigorous rounds of assessment to prove their worth. After which they invested a large part of their time researching and compiling the most relevant data for our readers.

The editorial board has been involved in producing this book since its inception. They have spent rigorous hours researching and exploring the diverse topics which have resulted in the successful publishing of this book. They have passed on their knowledge of decades through this book. To expedite this challenging task, the publisher supported the team at every step. A small team of assistant editors was also appointed to further simplify the editing procedure and attain best results for the readers.

Apart from the editorial board, the designing team has also invested a significant amount of their time in understanding the subject and creating the most relevant covers. They scrutinized every image to scout for the most suitable representation of the subject and create an appropriate cover for the book.

The publishing team has been an ardent support to the editorial, designing and production team. Their endless efforts to recruit the best for this project, has resulted in the accomplishment of this book. They are a veteran in the field of academics and their pool of knowledge is as vast as their experience in printing. Their expertise and guidance has proved useful at every step. Their uncompromising quality standards have made this book an exceptional effort. Their encouragement from time to time has been an inspiration for everyone.

The publisher and the editorial board hope that this book will prove to be a valuable piece of knowledge for researchers, students, practitioners and scholars across the globe.

List of Contributors

S. Yashkina, V. Doroganov, E. Evtushenko, O. Gavshina and E. Sysa
Department of Technology of Glass and Ceramics, Belgorod State Technological University named after V.G. Shukhov, Belgorod, Russia

T. Shchemelinina and E. Anchugova
Laboratory of Biochemistry and Biotechnology, Institute of Biology, Komi Science Center, Ural Branch of the Russian Academy of Sciences, Syktyvkar, Russia

O. Kotova, D. Shushkov, G. Ignatyev and M. Markarova
Laboratory of Technology of Mineral Raw, Institute of Geology named after Academician N.P. Yushkin Komi Science Center of the Ural Branch of the Russian Academy of Sciences, Syktyvkar, Russia

A. Kokh, A. Kuznetsov, N. Kononova and V. Shevchenko
Sobolev Institute of Geology and Mineralogy SB RAS, Novosibirsk, Russia

K. Kokh
Sobolev Institute of Geology and Mineralogy SB RAS, Novosibirsk, Russia
Novosibirsk State University, Novosibirsk, Russia

B. Uralbekov and A. Bolatov
Al-Farabi Kazakh National University, Almaty, Kazakhstan

V. Svetlichnyi
Tomsk State University, Tomsk, Russia

M. Harja
Gheorghe Asachi Technical University of Iasi, Iai, Romania

A. Ponaryadov
Institute of Geology Komi SC UB RAS, Syktyvkar, Russia

S. Sun
Institute of Non-metallic Minerals, Department of Geological Engineering, School of Environment and Resource, Southwest University of Science and Technology, Mianyang, People's Republic of China

D. Kamashev
Institute of Geology Komi SC UB RAS, Syktyvkar, Russia

N. Min'ko and O. Dobrinskaya
Belgorod State Technological University named after V G Shukhov, Belgorod, Russia

E. Yatsenko, A. Ryabova and L. Klimova
South Russian State Polytechnical University (NPI), Novocherkassk, Russia

V. Ilyina
Institute of Geology KarRC RAS, Petrozavodsk, Russia

V. Strokova and D. Bondarenko
Department of Material Science and Material Technology, Belgorod State Technological University named after V.G. Shukhov, Belgorod, Russia

I. Moreva, O. Sysa and V. Bedina
Belgorod State Technological University named after V G Shukhov, Belgorod, Russia

O. Sysa, I. Moreva and V. Loktionov
Institute of Chemical Technology, BSTU named after V.G. Shukhov, Belgorod, Russia

V. Nelyubova, O. Masanin, S. Usikov and V. Babaev
Department of Materials Science and Materials Technology, Belgorod State Technological University named after V.G. Shukhov, Belgorod, Russia

T. Murtazaeva and M. Saidumov
Millionshchikov Grozny State Oil Technical University, Grozny, Russia

V. Hadisov
Millionshchikov Grozny State Oil Technical University, Grozny, Russia
Ibragimov Complex Research Institute, RAS, Grozny, Russia

V. Loganina, E. Mazhitov and V. Demyanova
Department "Quality Management and Technology of Construction Production", Penza State University of Architecture and Construction, Penza, Russia

N. Kozhukhova
Belgorod State Technological University named after V.G. Shukhov, Belgorod, Russia

K. Sobolev
University of Wisconsin-Milwaukee, Milwaukee, USA

M. Elistratkin, V. Lesovik and N. Chernysheva
Department of Building Materials, Products and Designs, Belgorod State Technological University named after V.G. Shukhov, Belgorod, Russia

E. Glagolev
Department of Construction and Municipal Economy, Belgorod State Technological University named after V.G. Shukhov, Belgorod, Russia

P. Hardaev
Department Industrial and Civil Engineering, East Siberian State University of Technology and Management, Ulan-Ude, Russia

A. Balykov, T. Nizina and D. Korovkin
Department of Building Structures, Ogarev Mordovia State University, Saransk, Russia

M. Salamanova
Grozny State Oil Technical University named after Academician M.D. Millionshchikov, Grozny, Russia
Complex Research Institute named after H.I. Ibragimov, Russian Academy of Sciences, Grozny, Russia

S.-A. Murtazaev
Grozny State Oil Technical University named after Academician M.D. Millionshchikov, Grozny, Russia
Complex Research Institute named after H.I. Ibragimov, Russian Academy of Sciences, Grozny, Russia
Academy of Sciences of the Chechen Republic, Grozny, Russia

A. Alashanov and Z. Ismailova
Grozny State Oil Technical University named after Academician M.D. Millionshchikov, Grozny, Russia

I. Zhernovsky
Belgorod State Technological University named after V.G. Shukhov, Belgorod, Russia

A. Tolstoy, E. Glagolev and L. Zagorodniuk
Belgorod State Technological University named after V.G. Shukhov, Belgorod, Russia

Y. Ogurtsova, E. Gubareva, M. Labuzova
Department of Materials Science and Technology, Belgorod State Technological University named after V.G. Shoukhov, Belgorod, Russia

M. Khubaev
Millionshchikov Grozny State Oil Technical University, Grozny, Russia

G. Le Saout and J.-C. Roux
C2MA, IMT Mines Ales, Univ Montpellier, Ales, France

R. Idir
CEREMA, DIM Project Team, Provins, France

M. Garkavi, A. Artamonov, A. Pursheva and M. Akhmetzyanova
JSC "Ural-Omega", Magnitogorsk, Russia

E. Kolodezhnaya
Department of Mining and Ecology, Institute of Comprehensive Exploitation of Mineral Resources, Russian Academy of Sciences, Moscow, Russia

C. Sequeira
GeoBioTec, Geosciences Department, University of Aveiro, 3810-193 Aveiro, Portugal

T. Kuladzhi
Lomonosov North (Arctic) Federal University, Arkhangelsk, Russia

S. Aliev and M. Hubaev
Millionshchikov Grozny State Oil Technical University, Grozny, Russia

D. Mishin and S. Kovalev
Department of Technology of Cement and Composite Materials, Institute of Chemical Technology, Belgorod State Technological University named after V.G. Shukhov, Belgorod, Russia

S. Titov and A. Kazakov
Federal State Budgetary Educational Institution, Russian University of Transport (MIIT), Moscow, Russia

V. Nelyubova, Yu. Ogurtsova and M. Rykunova
Belgorod State Technological University named after V.G. Shukhov, Belgorod, Russia

M. Saydumov
Millionshchikov Grozny State Oil Technical University, Grozny, Russia

M. Nakhaev
Chechen State University, Grozny, Russia

V. Lesovik, A. Volodchenko and I. Lashina
Belgorod State Technological University Named After V.G. Shukhov, Belgorod, Russia

H.-B. Fischer
Bauhaus-Universität Weimar, Weimar, Germany

D. Arduin
GeoBioTec, Geosciences Department, University of Aveiro, 3810-193 Aveiro, Portugal

A. Velosa
GeoBioTec, Geosciences Department, University of Aveiro, 3810-193 Aveiro, Portugal
RISCO, Civil Engineering Department, University of Aveiro, 3810-193 Aveiro, Portugal

V. Konovalov, A. Fedorov and A. Goncharov
Belgorod State Technological University named after V.G. Shukhov, Belgorod, Russia

A. Volodchenko
Belgorod State Technological University named after V.G. Shukhov, Belgorod, Russia

T. Nizina, A. Balykov and V. Volodin
Department of Building Structures, Ogarev Mordovia State University, Saransk, Russia

V. Galdina, E. Gurova, P. Deryabin, M. Rashchupkina and I. Chulkova
Department of Building Structures, Ogarev Mordovia State University, Saransk, Russia

W. Hajjaji, S. Andrejkovičová and A. Cerqueira
GeoBioTec, Geosciences Department, University of Aveiro, 3810-193 Aveiro, Portugal

S. Kovalyov
Department of Technology of Cement and Composite Materials, Chemical Technology Institute, Belgorod State Technological University named after V.G. Shukhov, Belgorod, Russia

B. Mu
Key Laboratory of Clay Mineral Applied Research of Gansu Province, Center of Eco-material and Green Chemistry, Lanzhou Institute of Chemical Physics, Chinese Academy of Sciences, Lanzhou 730000, People's Republic of China

A. Zhang
Key Laboratory of Clay Mineral Applied Research of Gansu Province, Center of Eco-material and Green Chemistry, Lanzhou Institute of Chemical Physics, Chinese Academy of Sciences, Lanzhou 730000, People's Republic of China
Center of Materials Science and Optoelectronics Engineering, University of Chinese Academy of Sciences, Beijing 100049, People's Republic of China

Y. Kang, A. Hui
Key Laboratory of Clay Mineral Applied Research of Gansu Province, Center of Eco-Material and Green Chemistry, Lanzhou Institute of Chemical Physics, Chinese Academy of Sciences, Lanzhou, China

A. Titov
National Research University, Novosibirsk, Russia
Sobolev V.S. Institute of Geology and Mineralogy of SB RAS, Novosibirsk, Russia

V. Zaikovskii
National Research University, Novosibirsk, Russia

P. M. Larionov
National Research University, Novosibirsk, Russia
Boreskov Institute of Catalysis of SB RAS, Novosibirsk, Russia

A. Izatulina and O. Frank-Kamenetskaya
Department of Crystallography, St. Petersburg State University, St. Petersburg, Russia

M. Zelenskaya
Department of Botany, St. Petersburg State University, St. Petersburg, Russia

S. Lipko and V. Tauson
Vinogradov Institute of Geochemistry SB RAS, Irkutsk, Russia

I. Lipko and K. Arsent'ev
Limnological Institute SB RAS, Irkutsk, Russia

F. Wang
Key Laboratory of Clay Mineral Applied Research of Gansu Province, Center of Eco-Material and Green Chemistry, Lanzhou Institute of Chemical Physics, Chinese Academy of Sciences, Lanzhou, People's Republic of China
College of Petroleum and Chemical Engineering, Qinzhou University, Qinzhou, People's Republic of China

Y. Zhu, W. Wang and A. Wang
Key Laboratory of Clay Mineral Applied Research of Gansu Province, Center of Eco-Material and Green Chemistry, Lanzhou Institute of Chemical Physics, Chinese Academy of Sciences, Lanzhou, People's Republic of China

L. Yarg, I. Fomenko and D. Gorobtsov
Department of Engineering Geology, Russian State Geological Prospecting University (MGRI), Moscow, Russia

E. Levchenko, I. Spiridonov and D. Klyucharev
FSBI IMGRE, Moscow, Russia

M. Abou Zahr Diaz, M. A. Alawiyeh and M. Ghaboura
Department of Mineral Developing and Oil & Gas Engineering, Engineering Academy, RUDN University, Moscow, Russia

F. Dong, X. Zhao, Q. Dai, Q. Li, Y. Luo and S. Deng
School of Environment and Resource, Southwest University of Science and Technology, Mianyang, China

K. Kuroda, K. Toda and Y. Kobayashi
Graduate School of Engineering, Hokkaido University, Hokkaido, Japan

H. Kamegamori and K. Lawrence
Graduate School of Engineering, Hokkaido University, Sapporo, Japan

T. Sato and T. Otake
Faculty of Engineering, Hokkaido University, Sapporo, Japan

L. Z. Zhang
Department of Mineral Developing and Oil & Gas Engineering, RUDN University, Moscow, Russia
Liaoning Shihua University, Fushun, China

H. Y. Sun
Department of Mineral Developing and Oil & Gas Engineering, RUDN University, Moscow, Russia
Qinhuangdao Experimental Middle School, Qinhuangdao, China

I. Shadrunova and A. Proshlyakov
Academic N.V. Melnikov Institute of Problems of Comprehensive Exploitation of Mineral Resources, Russian Academy of Sciences, Moscow, Russia

S. Krivovichev
Kola Science Center, Russian Academy of Sciences, Apatity, Russia
Department of Crystallography, St. Petersburg State University, St. Petersburg, Russia

S. Yudintsev
Institute of Geology of Ore Deposits, Petrography, Mineralogy, and Geochemistry, Russian Academy of Sciences, Moscow, Russia

A. Pakhomova
Deutsches Elektronen-Synchrotron (DESY), Petra III, Hamburg, Germany

S. Stefanovsky
Frumkin Institute of Physical Chemistry and Electrochemistry, Russian Academy of Sciences, Moscow, Russia

E. M. Tanoh Boguy and T. Chekushina
Department of Mineral Developing and Oil & Gas Engineering, Engineering Academy, RUDN University, Moscow, Russia

O. Frank-Kamenetskaya, D. Vlasov and M. Zelenskaya
St. Petersburg State University, St. Petersburg, Russia

V. Rytikova and V. Manurtdinova
State Museum of Urban Sculpture, St. Petersburg, Russia

V. Parfenov
St. Petersburg Electrotechnical University «LETI», St. Petersburg, Russia

N. Borovkova
Mining Museum, St. Petersburg Mining University, St. Petersburg, Russia

M. Machevariani
Assistant of the Department of Mineralogy, Crystallography and Petrography, St. Petersburg Mining University, St. Petersburg, Russia

S. Moutinho and J. Nobre
RISCO, Civil Engineering Department, University of Aveiro, 3810-193 Aveiro, Portugal

C. Costa, A. Velosa and F. Rocha
RISCO, Civil Engineering Department, University of Aveiro, 3810-193 Aveiro, Portugal
GeoBioTec, Geosciences Department, University of Aveiro, 3810-193 Aveiro, Portugal

Â. Cerqueira, C. Sequeira, D. Terroso and P. Morgado
GeoBioTec, Geosciences Department, University of Aveiro, 3810-193 Aveiro, Portugal

Index

Printed in the USA
CPSIA information can be obtained
at www.ICGtesting.com
JSHW011400091023
49903JS00004B/37